Praise for *My Regenerative Kitchen*

"Camilla has always been a fearless kitchen warrior, industry advocate, and guiding force for all things edible. Her commitment to effecting change in our faltering food system is infectious, and her vision for how we all can be part of the solution through small but vital pivots at home is empowering. With *My Regenerative Kitchen*, Camilla leads us with passion and confidence toward a brighter, healthier, and more delicious future for our families and our planet."

—**Gail Simmons**, food expert;
TV host; author of *Bringing It Home*

"Camilla Marcus is one of my heroes, and *My Regenerative Kitchen* proves why. If it isn't regenerative, we shouldn't do it. If it's wasteful, thoughtless, and doesn't take our climate crisis into account, we just shouldn't do it. As importantly, in the kitchen, you don't have to sacrifice pleasure, deliciousness, fun, and creativity when you're cooking. That intersection is Camilla's and is what this superb new book beautifully illustrates. In twenty years, this will be a classic of its time as much for its recipes as its ideas, and you don't have to wait until then to make this book a cornerstone of your sustainable lifestyle."

—**Andrew Zimmern**,
Emmy-winning and four-time
James Beard Award–winning TV personality;
chef, writer, and social justice activist

"Camilla Marcus has long been at the forefront of conversations related to sustainability. In this brilliant and beautiful book, she lays out a means of achieving such sustainability in our home kitchens and in our lives—moving the conversation from the realm of high-end, well-funded restaurant kitchens and putting it within reach of all of us. This is an important work—one that is more than just a clarion call for needed change, but also a well-articulated roadmap of how we can all participate in the solutions."

—**Chris Kostow**, executive chef,
The Restaurant at Meadowood

"Camilla is a star, and *My Regenerative Kitchen* is just her latest brilliant contribution. This book is insightful and fun, and she continues to show that innovation can be delicious!"

—**Wes Moore**, governor of Maryland

"The pages of *My Regenerative Kitchen* draw us into Camilla's world, where what's good for the Earth is the key to what's good for our health. Camilla is a fresh new voice and a guide for our culture through the regenerative movement. Her book transports you into her home, where you won't want to leave until you have soaked up all her goodness.

—**Sarah Harrelson**,
editor-in-chief, *Cultured*

"There is no one I trust more implicitly than I do Camilla Marcus when it comes to learning as much as I can about food, where it comes from, and how it gets to our plates—and, for the sensual purpose of consumption, how to make it taste as good as possible. This book is the two-in-one that serves as a course in food sourcing and . . . a course in eating! If you only buy one cookbook, let it be this one."

—**Leandra Medine Cohen**, author and blogger

"Camilla Marcus's new cookbook combines creativity and environmental consciousness to provide a profoundly delicious group of recipes to nourish ourselves and the planet."

—**Dana Cowin**, author; former editor-in-chief, *Food & Wine*

"Camilla has been a pioneer in the regenerative food space, bringing to life the most gorgeous and thoughtful meals I have ever experienced. Beyond that, she is the ultimate hostess, making the sustainable dining experience feel accessible and worthwhile for anyone."

—**Jaclyn Johnson**, founder, Create & Cultivate

"Camilla Marcus is a force of nature—a chef, mother, activist, entrepreneur, consummate host, and someone I'm lucky to call a close confidante and friend. I'm endlessly inspired by her vision and conviction in working to create a better, more regenerative world for all of us. To be in Camilla's orbit and sit at her table is a blessing. I'm so excited for the world to finally experience her magic through her first cookbook."

—**Jing Gao**, chef and author of James Beard Award–winning *The Book of Sichuan Chili Crisp*

"*My Regenerative Kitchen* reminds us of the impact our practices have on the planet, and its recipes are delicious tools to do better."

—**Clare de Boer**, four-time James Beard Award–nominated chef, Stissing House

"A magnificent first cookbook by Camilla Marcus—*My Regenerative Kitchen* is pure nourishment of the appetite, soul, and spirit."

—**Michael Solomonov**, chef and co-owner, Zahav

"Camilla's beautiful book highlights how interconnected we all are—to each other, to our resources, and to the future. Her considered perspective and delicious recipes inspire me in my own cooking."

—**Eunice Byun**, cofounder, Material Kitchen

"No matter where you are in life, protecting and reveling in what you're passionate about is all there is. Read this book and allow Camilla to inspire you in a myriad of beautiful ways to do just that."

—**Rich Torrisi**, chef, Sadelle's

"Camilla is one of the most thoughtful women in the restaurant industry and is always putting people and the planet first. I am thrilled to see her first cookbook out in the world and hope readers are just as inspired as I am to be more mindful about what goes in (and out!) of our home kitchens."

—**Kerry Diamond**, founder and editor, *Cherry Bombe*

My Regenerative Kitchen

My Regenerative Kitchen

PLANT-BASED RECIPES AND
SUSTAINABLE PRACTICES
TO NOURISH OURSELVES
AND THE PLANET

Camilla Marcus

FOREWORD BY ALICE WATERS
PHOTOGRAPHY BY BEN ROSSER

Chelsea Green Publishing
White River Junction, Vermont
London, UK

Editor: Rebecca Springer
Copy Editor: Karen Wise
Proofreader: Rachel Markowitz
Indexer: Linda Hallinger
Designer: Melissa Jacobson

Printed in the United States of America.
First printing October 2024.
10 9 8 7 6 5 4 3 2 1   24 25 26 27 28

Our Commitment to Green Publishing

Chelsea Green sees publishing as a tool for cultural change and ecological stewardship. We strive to align our book manufacturing practices with our editorial mission and to reduce the impact of our business enterprise in the environment. We print our books using vegetable-based inks whenever possible. This book may cost slightly more because it was printed on paper from responsibly managed forests, and we hope you'll agree that it's worth it. *My Regenerative Kitchen* was printed on paper supplied by Versa that is certified by the Forest Stewardship Council.®

Library of Congress Cataloging-in-Publication Data

Names: Marcus, Camilla, 1985– author. | Waters, Alice, writer of foreword. | Rosser, Ben, photographer.
Title: My regenerative kitchen : plant-based recipes and sustainable practices to nourish ourselves and the planet / Camilla Marcus ; foreword by Alice Waters ; photography by Ben Rosser.
Description: White River Junction, Vermont ; London, UK : Chelsea Green Publishing, [2024]
    | Includes bibliographical references and index. |
Identifiers: LCCN 2024025341 | ISBN 9781645022541 (hardcover) | ISBN 9781645022558 (ebook)
Subjects: LCSH: Vegan cooking. | LCGFT: Cookbooks.
Classification: LCC TX837 .M299 2024 | DDC 641.5/6362—dc23/eng/20240724
LC record available at https://lccn.loc.gov/2024025341

Chelsea Green Publishing
White River Junction, Vermont, USA
London, UK
www.chelseagreen.com

To my four under forty, for their future.

Contents

Cultivating Hope in the Soil

In the urgent quest to address climate change and safeguard our planet's future, there exists a profound yet often overlooked solution—one that lies right beneath our feet. Within the realm of regenerative agriculture, we unearth a paradigm-shifting approach that not only promises to restore our environment but also to allow us to thrive within it.

In simple yet profound ways, the choices we make about how we eat can help us—and the communities in which we live—lead healthier and more joyful lives. In the pages of *My Regenerative Kitchen*, Camilla Marcus shows us how eco-conscious cooking is synonymous with the vision of regenerative organic farming and its focus on nurturing healthy soils to protect our planet. Like farmers, chefs can be environmental stewards. Camilla helps us understand how to waste less and eat with earth-friendly practices in mind, nourishing ourselves with beautiful, organic, fresh dishes.

Regenerative organic farming, with its focus on nurturing the soil as a vibrant and diverse ecosystem, holds the key to our survival. It is a technology born from the wisdom of nature, refined through rigorous scientific inquiry, and poised for widespread adoption. At its core, regenerative agriculture harnesses the innate capacity of soil to sequester and hold carbon. By treating soil not merely as dirt but as a living, breathing entity teeming with life, we unlock its transformative potential. This revolutionary approach not only mitigates the escalating levels of atmospheric carbon but also enhances the resilience of our lands against the ravages of climate extremes.

Yet the impact of regenerative agriculture extends far beyond its role as a climate savior. With enriched soil carbon comes a flourishing soil biology—a bustling ecosystem brimming with biodiversity. This abundance not only sustains the health of our ecosystems but also enriches the nutritional value of our food.

Education and food are two universal rights. Everyone deserves to eat nourishing food. And everyone goes to school (or at least ought to), and public education has the potential to reach every person on the planet. This proposition would not only bring money to local communities but would bring the essential values of stewardship, nourishment, interconnectedness, diversity, and, I dare say, democracy to the next generation, directly through the cafeteria doors. It would activate students every single day in the way they yearn for, giving them an instant connection to each other at the table and, through the ingredients on their plate, to the greater world.

I can say all of this with conviction because I have seen firsthand that it is possible to create a program that rethinks food and academics from the ground up and have it widely adopted. Thirty years ago, I started the Edible Schoolyard Project at Martin Luther King Jr. Middle School in Berkeley, California, by creating garden and kitchen classrooms to teach all academic subjects to students. Because of the human values this project celebrated—stewardship of the land, nourishment, community, diversity, beauty—the Edible Schoolyard Project has inspired a network of over 6,200 like-minded programs around the world.

In 2018, Camilla unveiled west~bourne, a groundbreaking zero-waste dining spot in the heart of New York City, earning praise not only for her delectable dishes but also for her mindful, forward-looking, community-driven ethos. Camilla emerged as a visionary, guiding a venture that upholds ethics, innovation, and local involvement, all while championing the cause of farmers and environmental stewardship.

We have to think completely differently about our food system—about the relationship we have with our producers and the responsibility we owe each other. It will take a total paradigm shift. In essence, regenerative agriculture offers us a profound opportunity to recalibrate our relationship with the Earth. It invites us to reimagine our role as stewards of the land, working in harmony with nature's rhythms rather than against them. By harnessing the surplus carbon in our atmosphere, we can breathe new life into our soils, cultivating a future of abundance and vitality.

As we stand at the precipice of unprecedented environmental challenges, let us heed the call of regenerative agriculture. Let us embrace the soil as a sacred trust, a reservoir of hope from which the seeds of a sustainable future shall sprout. For in the embrace of healthy soil lies our greatest salvation—a beacon of resilience illuminating the path toward a thriving planet for generations to come.

—ALICE WATERS

Welcome to a Work in Progress

"We need to work with nature, not against it."
—David Attenborough

I am an activist at heart. When I was a kid, I wrote letter after letter (with the help of my parents) to my local representatives, to members of Congress, and even to the president, urging consideration and changed policies with regard to environmental protection. No one told me at the age of nine that those efforts were useless or that my letters would likely go unread, and fortunately no one made me feel too small to raise my voice about an issue I cared about. In fact, my parents framed the generic, stock letters the government sent back, so that I would be proud to have spoken up.

Today when I hear the word "activist," it's often politicized and considered loaded. I've been asked to pull it from my bio on more than one occasion. Yet, at its heart, being an activist is about being *active*. Being willing to take a step diverging from the status quo, toward something that makes our collective lives better. Believing that individual actions can shift a paradigm and in turn influence and move people to do the same, creating a collective groundswell. My work is rooted in my activist mission, and the audacity to advocate for change runs through my veins.

I'm the youngest child and only daughter of a large family, born and raised as a rare second-generation Los Angeles native. My family came from New York by way of the Midwest and then all the way to the Pacific Ocean. Moving west was about opportunity and a chance to live a life in balance with the environment. My family history is rooted in life science, and the future of human health is a constant conversation in our homes. Wellness for ourselves and the elements around us is the air we breathe. Living by the water amidst suburban sprawl, I was born with an openness to and craving for risk, and with much less expectation than that placed

on my two older brothers. I was free to be a constant explorer who was never done learning, at home in this newish American city with salty air that seems to remain crisp with the constant promise of a fresh start.

Although my family wasn't anchored in cooking, food somehow became my primary love language, and I eagerly sought inspiration anywhere I could find it. I am drawn to food as one of our greatest universal languages, one that spans generations and cultures. From a very young age I begged to try just about every restaurant I could, much to the surprise of my parents, who indulged my quest despite being perplexed by it. A lot has been said about the food scene in LA, but being raised in the heart of it, I've always found the food community and culture to be layered, nuanced, and diverse: from Koreatown to taco trucks, to dim sum in the San Gabriel Valley, to hidden Italian gems, to sushi and kaiseki in Little Tokyo, and everything in between. Many restaurants that have stood the test of time in our town are family-run operations with a care for quality and intimacy. California grows over a third of the country's vegetables and nearly three-quarters of the country's fruits and nuts, and a deep reverence for

ingredients is one of the hallmarks of cooking here. This is the terroir of
this book's ethos and a tribute to the bounty of California's farms.

The vast bounty of the West has instilled my innate sense of gratitude
and awe for Mother Nature—inherently wild, rhythmically unpredictable, and magnificently monumental. Living with the land, not just on it,
is in my Californian nature. Being a human of the Earth—to listen, adapt,
and grow to live in balance with the planet as it ebbs and flows. I am
someone who could not bear to live a life of routine, so I thrive on the
unexpected qualities of the Earth, where you're never quite sure what
you'll get. That spontaneity is a gift rather than a limitation. The core
beauty of life lies in the friction, those moments when discomfort and
uncertainty spark an indescribable energy. We humans are flawed, imperfect, and constantly evolving, despite our urges to smooth it all out
uniformly. I live for that rush. I thrive on not knowing for certain what I
might find at a farmers' market, and I revel in the thrill of being creative
and flexible, making substitutions or shifts in how I cook based on what
I might have on hand or can procure from local growers.

Cooking is about letting go and living with abandon. Finding your way with what's grown and releasing the instinct to exert control. It's the very feeling I imagine musicians vibe on when playing jazz, not knowing where the notes and rhythm will lead or end up, which is entirely the point of the experience. To be present, savoring the winding road of the process. While improvisational cooking might feel daunting at first, as is so often the case when we depart from routine, give yourself some grace and surrender to experimenting. With time and some trial and error, hopefully you too will find that sweet spot where you can find joy in adapting alongside nature as she wants and discover that not being in control can be both liberating and inspiring.

When I was in culinary school at the French Culinary Institute, we had the opportunity to operate a restaurant, L'École, as part of the curriculum. This was my first experience in a professional kitchen. I loved every aspect of the grit, intensity, smells, heat, and eclectic brigade. I reveled in the electric energy of chaos and order emanating from that environment. The vegetarian station instantly became my favorite rotation, despite being all but ignored by the rest of our crew. Most dismissed it or scowled when they were assigned to the role. Yet it was the one place where the instructors and head chefs did not hand out recipes to follow. For those brief nights of prep and service that I had my vegetarian domain, I was free from the requirements of conformity, of repetition, and of being told what to do. The catch was I didn't get to pick what I had to work with, as the station was largely stocked with whatever was left over from the various teaching kitchens that week. One person's trash, another's treasure defined. No scripts and scraps—I was in heaven. I had to dig deep and find ingenuity in what others discarded or overlooked.

Our core curriculum predominantly centered on protein and butter—as does French cuisine. It was the experience of a lifetime to learn about new technology and to make a turducken on Thanksgiving with Dave Arnold and garnish scratch-made tarts with precious grace alongside Jacques Pépin. While it certainly expanded my palate in ways I never thought possible and taught me invaluable techniques, I couldn't shake the feeling that it was cheating a bit, that it was the approach of a different era that didn't seem to reflect my current reality. Of course a knob of butter will make a steak delicious. It was excessive and maybe sexy to an extent, yet it felt old-school and rigid, not to mention divorced from the imminent threats to our food system.

The dissonance of recipes that hadn't changed in decades, if not longer, fell flat for me. Where was the ingenuity to learn ways to advance the food system, to feed our planet, and to nurture our health? To bring the power of food into the twenty-first century, where we need food to transform for our climate? That window cracked open to me at the start of my career and has fueled my passion ever since.

When I began my career in food and cooking, there were hardly any notable vegetarian restaurants, and the best-regarded chefs were not known for their environmental perspectives. Lavish events had to have steak, seafood, and caviar. Food magazines did not include plant-based meals, let alone a single dish that wasn't a side in their holiday issues. Vegetables were not at center stage or in high regard. Fast-forward to the 2020s, and Sophia Roe and Marcus Samuelsson prepared an entirely vegan dinner for the Met Gala. Amanda Cohen's tour de force Dirt Candy has redefined fine dining. And Hetty Lui McKinnon was the first to do an entirely vegetarian Thanksgiving spread for the coveted *Bon Appétit* holiday print issue in 2022. Just because it's how it may have been does not mean it's how it has to or will be. We can flip the script or dare to run without a script at all.

In this book, as with the ethos of my recipes and culinary style, I chose film photography to capture everything just as it is, raw and unedited reality in all its messiness—just like Mother Nature. We wanted to honor the slowness of an intentional process, as I believe consistency and immediacy are overrated. With film, you never know exactly what will

be, which can illuminate more than you can imagine. *Res ipsa loquitur*—Latin for "the thing speaks for itself"—is my husband's mantra, expressing a core tenet of this book to give space and acceptance to let things be what they are. I picked a photographer, Ben Rosser, whom I've been collaborating with for over a decade but who had never shot a cookbook before. We had no idea how the shots would turn out, as we surrendered to the unpredictable power of the medium. We trusted one another and the process to receive and share something unique unto itself, unfiltered.

This book is about being so progressive that we go back in time, to before the Industrial Revolution, and find innovation in tradition, as nature intended. Ancient techniques and ingredients with a modern purpose and perspective. The data and reality are clear: Our topsoil has eroded, threatening our ability to grow food on productive land, our atmosphere is warming at an alarming rate, and we are running out of options for waste disposal. The climate crisis is here and looming. The good news is we can do something about it.

We make more decisions about what we eat and drink in our daily lives than almost anything else. Every single human has to eat and drink to survive and thrive. So, imagine if, little by little, those decisions shifted to be more intentional and aware of our impact. What if our daily choices were made through the perspective of how we could support climate solutions from our homes and kitchens? Release the draw to buy perfect-seeming produce year-round in plastic packages at our grocery stores. Reconsider cooking without regard for waste in search of a pristine preparation. Try new ingredients or change a menu based on what you can find in season. Radically give up control. Imagine the ripple effects. Consider how systems, policies, and our culture would transform if these small decisions every single day moved one by one, human by human, little by little, regularly, and consistent new behaviors compounded and grew to have a tremendous collective impact. It's not only possible, it's already happening.

You too can be a radical, by letting go of how it's been done. An agent of change—as a citizen, as a consumer, and as a cook. It just takes one step, followed by another. I'm not here to coax you into becoming a full-time vegan (unless you want to). I am here to give you about a hundred dishes to choose from to integrate more vegetarian and plant-based cooking into your routine and to bring a climate consciousness to how you source and prepare food in your home, whether personally or when hosting. These recipes are approachable and delicious, that I can assure you. Maybe, over time, they might shift from being your kitchen side hustle to a daily practice.

Through stories, photography, research, recipes, techniques, and pro tips, this book seeks to be an inspiring guide for your personal journey to radically let go and regenerate our Earth through how you live, cook, eat, and gather every day. When we honor our Earth, we nurture our own health—the true meaning of regeneration—finding harmony between our soil and soul. Small shifts—together—building a big impact. Welcome to the collective work in progress, ever evolving and always in pursuit.

To Begin, You Just Need to Take One Step

Our lives on Earth begin and end with our soil. It's the foundation for us all that lies beneath each of our feet and is the genesis of everything we grow to eat. We hear a lot now about carbon, but what does it really mean tangibly in your daily life? The journey of carbon emissions through food flows as follows, and some good news is that you can do something to impact each stage. As this book unfolds, we will follow this progression, from sourcing to cooking through disposal, so you can visualize how you contribute personally at each level of the carbon life cycle.

Sustainability Starts at the Source

Living in California, I've grown up with our growers, intimately connected to where my produce is cultivated and who is caring for the land to make it all possible. I plan my week around when I can go to my favorite local farmers' markets. There are farms that share their gems only on certain days in certain neighborhoods, so it's an exhilarating hunt to come upon what's the freshest and most unique of the season in that very—often fleeting—moment. It's a central gathering place across so many different communities in LA, where you run into just about everyone. We all converge to be humans, as one, together at the market, shedding our regular identities to just be people, finding and connecting with beautifully grown food.

Sourcing is where my cooking journey always begins. Fittingly, this is where we can have the single largest impact on the climate crisis—to recarbonize our soil. It's where we are losing the most, and for that reason have the broadest opportunity to reinvigorate. More than half of the carbon emissions of food products are derived from agriculture. Therefore, sustainability quite literally starts at the source, with the farms that grow our food. It's also the part of the food chain that we as consumers and cooks can have the most choice about. What this means in practice: Shop locally and from mindful farmers as directly as possible and be discerning about the products and brands you purchase at your grocery

stores. I'll be sharing every bit of knowledge I have, from regenerative farms across the country to specific regenerative brands that you can accessibly swap into your pantry.

Unsurprisingly, the carbon footprint of animal products far eclipses that of produce. While livestock is a piece of the regenerative agriculture principles to be discussed, this fact still vibrates highest for me, which is why there is no meat, fish, or poultry in this book. Transitioning toward a vegetarian if not entirely plant-based diet will have the most significant impact on the climate crisis. Many of us know this in theory but feel like this kind of change is an insurmountable mountain to climb. It's simple and obvious, just like Mom telling us to eat our veggies when we were young. Now more than ever, plants are the imperative. So let's have some fun with them.

Regeneration for the Next Generation

Food accounts for almost one-quarter of all greenhouse gas emissions globally. The *only* system that can capture carbon in time to address the looming climate crisis is food and agriculture. And it's also the only system that we, as everyday consumers and humans, can have a proactive effect on—a cosmic kismet.

We've long held to the myth of what the Industrial Revolution promised: that chemically driven monocrop farming is the only way to grow food at scale. That's what the lobbyists representing "big food" want you to believe as indisputable fact. What is irrefutable now, however, is that these practices have been stripping our invaluable, limited resource of land of its healthy soil and yielding food devoid of nutrition. By contrast, regenerative agriculture could draw down over half of that carbon into the soil in the coming decades—enough to compensate for all of the world's transportation-related emissions. Think about that. Regeneration is about rebelling against convention that has not served us.

As a mother of young children, I'm consumed with worry about what world we will pass onto them. I founded my company, west~bourne, to pioneer consumer packaged goods for the next generation. My personal and professional pursuits converge on a singular mission: to align personal and planetary health. At west~bourne, our ethos is "eat well. do better. gather often." We offer chef-driven snacks and pantry staples—all impeccably sourced, utilizing regenerative, sustainable, and organic ingredients—meant to elevate everyday cooking and serve as unique additions to any gathering.

Soil is connected to almost everything that humans do—the plants we grow, the food we eat, and how we, as a collective, can combat climate change. That's why supporting and encouraging regenerative agriculture—which promotes and improves soil health and biodiversity—is an undeniable way to solve the climate crisis.

We now know why it matters, but what *is* regenerative agriculture?

The term "regenerative" is still emerging, much like "organic" was a few decades ago. "Organic" is now in the zeitgeist, signaling a healthier option, regardless of whether the average person could recite the full and proper definition or if they understand the difference between organic and non-GMO. Organic practices are a component of regenerative agriculture, a stepping stone, though not the whole picture.

There isn't a full, succinct, widely agreed-upon definition of "regenerative," *yet*. But there are a few guiding principles that encapsulate regenerative agriculture and that this book seeks to infuse across what it means to me to live regeneratively. If you want to dig even deeper, there are numerous leaders and organizations with incredible resources, like the Regenerative Organic Certified community, Kiss the Ground, and Mad Agriculture—all working throughout the cycle to support the transition of farms to regenerative practices. Broadly speaking, regenerative agriculture is a dynamic food ecosystem that seeks to achieve six primary goals:

1. restore soil health
2. prevent erosion (for example, through cover crops)
3. promote biodiversity
4. minimize waste
5. protect water quality and improve soil's water-holding capacity
6. reduce or eliminate use of chemical fertilizers, herbicides, and pesticides

An abridged version of what this looks like for a farm is as follows. No-till or reduced tilling techniques are used to minimize land degradation, further assisted by managed grazing. The land should not remain fallow, which means year-round planting of cover crops, double planting, and relay planting, where the next season is seeded while the current one is in harvest. All these techniques are designed to build climate resilience in the soil as the weather fluctuates. With an eye toward biodiversity, methods like crop rotation, alley cropping, and interseeding can naturally limit pests and nurture beneficial microbes, in turn reducing reliance on chemical pesticides. The result is more vibrant, nutrient-dense, and chemical-free food that's rehabilitating our land to support long-term sustainability and viability for farms and those who tend to them. It seems like the most obvious and fruitful win-win that exists. It does require time, capital, patience, and some optimism ("blue-sky thinking," as I call it), which are all within our reach and grasp.

It's important to note that the very purpose of regenerative agriculture—to redress the monocrop fallacy of farmland that's been stripping our land of nutrients, viability, and carbon—means that these practices must be individualized. There are some critics who need certifications and guidelines to endorse the transition to regenerative farming. However, if nature must take the lead, then it follows that different regenerative practices will complement different regions, and the specific conditions of each individual farm will require a different approach. One size can't and shouldn't fit all.

As we look to our future for the next generation, we will have to grow more food, using less land, and with less environmental impact. The idea that good farming and good food must converge is not news. Exactly a decade ago, in his book *The Third Plate: Field Notes on the Future of Food*, Dan Barber laid out a history following the Industrial Revolution and this new regenerative path in depth. His work was a call to arms for chefs and diners alike to usher in a new era of dining. Yet, here we are in 2024, and regenerative agriculture and cooking are still considered niche. I've even heard some herald them as the ultimate luxury or criticize them as a benefit exclusive to the 1 percent—both notions I strongly disagree with.

What will tip this movement to the mainstream? Sustainable agriculture leads to increased nutrient density in foods, which supports healthier microbiomes for us and the soil. By definition, what's better for the Earth is better for our health. And time to reverse the climate crisis is ticking down, so the moment is now. We don't have the luxury to keep this niche. We often look to famous chefs and notable restaurants to indicate what the next wave is or should be in food, with the rest following their lead. What if you, reading and cooking from this book, take the lead instead? Think of yourselves as a sphere of influence that collectively can shift the status quo by making different choices about how you buy, cook, and share food. My aim is to make regenerative living accessible. You can make regenerative food the next frontier, bringing the whole farm into your home on your table. Wouldn't that be radical?

As my dear friend Evan Marks of the Ecology Center, a Regenerative Organic Certified farm in Southern California, says, "It's time we bring in the *new era of culture*, the most important part of agriculture."

Of Soil and Soul

"Growing your own food is like printing your own money."
—RON FINLEY

A self-proclaimed "rebel with a green thumb," Ron Finley revolutionized South Central Los Angeles by converting desolate urban plots into edible gardens. As an LA native, I have long been inspired by his work. He too believes nature and its enriching bounty should not be reserved for the privileged few. He set roots to ensure that food deserts, lacking affordable access to fresh and healthy food, can indeed be transformed into food paradises. "Grow" doesn't have to be a four-letter word. His legacy proves that planting something, anything, wherever you are is possible, affordable, and powerful. He's become a friend since I moved back home, and he knows that I care more about what he thinks than almost anyone else. He doesn't mince words, and his passion for democratizing climate focus and food access knows no bounds. He doesn't believe in the word "can't"; nor do I. "Can't" is the limited thinking that got us into this environmental mess in the first place. It's what the incumbent industries want you to think. Ron hands it to you straight, and he's a thoughtful resource, particularly when it comes to biodiversity and practical solutions that individuals can activate.

For those looking to learn more about a climate-centered garden, Acadia Tucker is another force in this field. Her book *Growing Good Food: A Citizen's Guide to Backyard Farming* outlines step-by-step how an amateur grower can get started and sustain a regenerative home garden. Her techniques and plans help support carbon farming for every citizen, everywhere. Gardening is indeed an act of civic duty—one that we all can and must be practicing.

Plant something. You can grow food whether you live in a city apartment, on a sprawling plot of land, or in any kind of home. It will change how you look at food and inspire how you shop at the farmers' market or grocery store. You can start with windowsill herbs or a small potted plant like a blueberry bush. Raised garden beds or vertical garden towers are also great starter kits to get you going. If you're landscaping, think about perennials or try cover crops like buckwheat and rye, which are as replenishing as they are decorative. It's also a way to explore biodiversity and see for yourself the difference in taste and in soil health. Homegrown crops are a resource to inspire experimentation in the kitchen, and hunting for seeds is a rewarding pastime in and of itself; Row 7 Seeds has a thoughtful selection. I personally love reaching out to our local farms, like Windrose Farm or the Ecology Center, to see what seeds they are selling each season. Seed banks are storage facilities run by governments, nonprofits, and groups that preserve and protect seeds that may be threatened by invasive species, monoculture crops, or natural disasters, serving as another resource for cultural and culinary preservation.

People, no matter what their age, can learn new patterns with a spirit open to all possibilities. I've seen this come to life with my own family. Our young toddlers gain a sense of independence and responsibility as they tend to our garden. As Ron Finley says, children learn what they live. Our kids constantly try what we grow, which exposes them to new foods on a rotating basis. We talk about how it grows, why it grows, and what special qualities and vitamins each crop has and how that impacts our bodies. Every season we plant something new, too, to see how it evolves. Our kids see the time and effort it takes before a plant gets to harvest. They also learn to relinquish their favorites, season after season, with a humble understanding that everything has its unique place and time. At any phase of life, we can tap into that sense of wonder for the beneficence of nature.

Nature First, People Always

Everything is about people. Humans are what make our world as enriching as it is complex. Just as we are all stewards of this Earth, I wanted to take a moment to especially honor our growers and farmers: the true heroes who do the hard work, day in and day out, to cultivate and safeguard what nature creates. Through sharing their dedication and gifts with the Earth, they are the great translators, bringing cherished food into our homes. Many here have been teachers to me on my journey, and others I remain a fan of from afar. My hope is that you will find your way to one or all of them in some way or another. There will never be sufficient acknowledgment or gratitude for their sacrifice, dedication, and service to the land and to us as a community.

Cover Crops: The Unsung Superheroes

Cover crops are planted to cover the soil and protect it through the seasons instead of being harvested. Cover crops help manage soil erosion, soil fertility, soil quality, water, weeds, pests, diseases, biodiversity, and wildlife. Think of them as natural mulch, in contrast to the primary crops farms sell for profit. They are the unsung heroes making regenerative farms work. So, what if we did create a market for them?

If we start using these crops en masse, increased demand will help make these farms and sustainable practices even more viable for the future. Plus, they are delicious and higher in nutritional value than many primary crops. Some examples of cover crops that can be easily integrated into your cooking include winter rye, oats, barley, buckwheat, lentils, winter peas, daikon radish, turnips, and alfalfa sprouts, to name a few. You'll see them crop up throughout the recipes in this book, and in particular in my baking. Cover crop grains like buckwheat, sorghum, and oats are ideal to support regenerative farm practices and are also vitamin rich and naturally gluten free. The further I progress into learning about

cover crops, the more excited I get about cooking with them and transforming what others think is possible.

When you go to select your rice, pasta, or flour, take a pause. Industrial wheat production has been stripping our soil. Try integrating cover crops and wheat alternatives into your everyday cooking. Use barley or amaranth from Wild Hive Farm instead of typical industrial white rice. Read the label on your pasta and search for regenerative brands like Edison Grainery or Flour + Water, made with domestic regenerative semolina flour. Try crackers from Moonshot Snacks. Buy your lentils from Rancho Gordo for a change. Integrate high-protein legumes like chickpeas and beans from Simpli. These all have greater flavor, are nurturing to the soil, and are nutrient dense for our health. There are so many growers and makers bridging this gap and making it easy to stock your pantry with carbon-conscious ingredients.

A Whole-Farm Frame of Mind

At its core, biodiversity is simply about variety, the very antithesis of the monocrop industrial complex. For regenerative farms to be the new standard, a core tenet is fostering a vibrant, multispecies habitat in which all members cross-pollinate, feed the soil, and naturally support one another's growth. Plants need a thriving, diverse community to flourish, just as humans do.

Moving back in time before monocrop industrial farming means we have to think dynamically about the food we have and eat at home. As you start to connect closer to your local farm community and growers, open the dialogue about what else they grow that might not be sold at the moment. Get curious. Ask them what they are most excited about coming out of the land or what seeds they are thinking about planting in the future. Maybe even ask if you can purchase seeds or starts from them to grow at home. You'll uncover an entire world you maybe didn't realize was percolating.

Another wonderful starting place is following Sam Rogers (@samloves themarket), who has an incredible Substack dedicated to novel ingredients and farming practices. It's like having an expert and an encyclopedia, who happens to be lovely and charming, in your inbox each week. Embracing a whole-farm mentality fosters surprise and delight in your cooking and supports the noble pursuit of the farmers nurturing biodiversity.

A whole-farm mindset means regularly trying something new. Restaurants have long held the place of first discovery, where diners can taste and uncover a new ingredient. My hope is that you too can step into that role in your own life. See what happens when you buy a new ingredient, something you've never even heard of, say, once a month to begin. Thanks to the internet and social media, there are endless resources for what to do with unfamiliar produce. Don't be afraid to try to work with it. There's no such thing as failure when it comes to cooking. It's about the process of trial and error, and trial again. You'll figure out how to do it, with time and a little humor.

Growth with food and cooking is all about learning and discovery. For me, it's as much about the enrichment of the journey as it is about the final product that ends up on a plate. I hope the recipes in this book will introduce ingredients you maybe haven't heard of or weren't sure what to do with, while giving some insight into why they are special, how they contribute to the ecosystem, and what nutritional benefits they can offer you. Throughout my many years as a chef, I still like to challenge myself, and I revel in seeing a fruit or vegetable wholly unknown to me. It's like the butterflies of a first date.

Your Home as Your Restaurant

Once we have sourced mindfully, it's time to explore how to apply a regenerative mindset to how you cook and prepare food. Gather only what's needed, use everything on hand, experiment bravely, adapt to constantly changing circumstances, and nourish others with passion.

Since my childhood, restaurants have had a magnetic pull on me—a siren call with a distinct high-frequency vibration. I find myself entranced by the juxtaposed chaos and order, intensity and fluidity, the noise and the moments of elegant silence. The swirl of contrasts brought together by an invisible string linking every person, each playing a distinct role, bonded as a formation in the space, experiencing the same moment. Behind it all is a coordinated symphony that's both messy and orchestrated by humans who live to take care of others, who take risks and work tirelessly to create moments that transcend our daily lives.

Like farmers, restaurants have played a significant role in translating what comes from our Earth to how humans perceive and enjoy food. Chefs made kale cool and have elevated vegetarian dining to Michelin Star status. Restaurants have long been the messengers from growers to diners, bringing new ingredients to public consciousness, inspiring new combinations, and expanding cuisines and culinary practices into the mainstream. What if every home joined that crusade?

Your home can be your haven—and your restaurant. Whether for your family or guests or even just for yourself, think of it as a theater for your intentions and purpose. It's about getting scrappy and thinking holistically.

Kitchen Confidence

For years before the pandemic, I owned a restaurant on Sullivan Street in New York City called west~bourne, the first certified zero-waste restaurant in the city. This was where I started to connect to regenerative farmers and see the entire food ecosystem through the lens of impact at every stage. My guiding principles were simple: We can always do better, and anyone can cook. And yes, I really do believe that. It's core to who we are and, evolutionarily, somewhere inside each of us, hardwired for our survival. We just have to have the openness to tap into it. Half of our team at the restaurant had never stepped into a professional kitchen before, and yet it never took as long as you might expect to get someone up to speed. They just had to believe it too.

Cooking, for me, is like music: You feel it and flow with it. To be fair, I can't sing a tune for anything and was never particularly gifted with an instrument. Yet innately, I hear a musical rhythm within, and that's how I feel about cooking. Despite what restaurant ratings and rankings will have you believe, to me cooking is the antithesis of perfection. The friction I think some are fearful of when they cook is actually—through another lens—the very bit of exhilaration that's key to the experience. What you make might not turn out as you'd imagined, but that's not really the point. I urge you to harness that trepidation into finding your flow state and embracing the unexpected that gets your heart racing, even a small leap, on that fine line between where you might flop or learn something new.

When I'm hired to do a big event or cook for some large gathering, most of the dishes I serve I've never road tested or made in full before. They are often experiments—a compilation of ideas, fragments of inspiration, ingredients I've been curious about, new techniques, or even just colors at times. I'm fortunate to have other incredibly talented humans and teams who've indulged this wild, unscripted approach to piece it all together on site, braving the uncharted alongside me on the line. Because, no matter what, we can always make something work on the fly.

So, whether you're a restaurant chef, a home cook, or an amateur, I hope you can take these recipes with a grain of salt, as you would to try a new method, ingredient, or dish with the excitement and naivete to just give it a whirl. We are all creators, and the liberating act of creation is the reward unto itself. Cooking is not perfect and shouldn't be in pursuit of perfection—far from it. As with nature, the perfection lies in the imperfection.

Take Stock and Prep

The goal of these techniques is to bring professional cooking practices to your kitchen at home. I've included pro tips to inspire flexibility. If you

understand the purpose behind a technique or a chosen ingredient, then substitutes and alternatives become more apparent and easier. You can do your food shopping and assess what you have on hand, and even if you can't match the ingredients list exactly, you can at least make some variation of the dish. With whole-produce methods and recipes that highlight how upcycled food can shine, the techniques and concepts seek to empower you to buy and prepare your food with a refreshed perspective. There's nothing worse than being beholden to a recipe or feeling that you've overshopped, with too much left over or ingredients you might not know what to do with. My goal is to give you multiple avenues to improvise, so if you can't find an ingredient, go rogue. Riff and play jazz with what I've set out.

When you dine in a restaurant, few menu items, if any, are made *à la minute*, in that moment. Service would become untenable. Preparation for components and dishes takes place in the hours, days, and sometimes weeks in advance, while factoring in how long things will keep fresh. These recipes provide building blocks and hopefully encourage you to

make some elements in advance, storing or freezing them to be used just when you need them, for more efficiency and to give yourself space and time to experiment. At home, if you have 10 minutes in the morning and can make the salad as the starter for dinner that night but not dress it, your evening will run that much more smoothly. Or, if you have 20 minutes on a Sunday to prep and freeze a big batch of simple dough, you can defrost it when hosting friends later in the week. With the dough premade, you can create a savory tart with vegetables for dinner or a sweet tart with fruit for dessert in minutes. Considering just a few steps ahead can make an immeasurable difference and can help avoid stress when the time comes to making and plating a meal.

Think big—and small. Building flavor through each component of a dish builds a progression, where each element unfolds as you savor it. Vegetable stock is something that flows throughout the book as an enhanced base compared to water. Every restaurant keeps house-made stock on hand at all times, on rotating production. I often hear from friends and those I've cooked for that making stock at home feels overwhelming or intimidating. I'm here to demystify that and hopefully inspire you to keep stock as a staple. I share a recipe for vegan dashi, and I also give a primer on vegetable stock, a key vessel for upcycling food scraps.

Every chef has very specific brands and sourcing for dry goods and pantry items, so I'm sharing mine. There is indeed a difference. On that note, you'll see these recipes are clean and nutrient forward, which begins with your base cooking oil. I believe strongly in healthy and effective oils, which is why avocado oil is used throughout. Avocado trees are often co-planted alongside coffee and chocolate, two of the highest-demand crops globally, as they provide shade and mutually beneficial ecosystems to one another. Moreover, avocado oil has a rich yet neutral flavor with a very high smoke point, well suited for both sweet and savory cooking, grilling, or baking. It's an all-around utility player and my personal staple.

You'll notice quite a few condiments and preserves for just that purpose too—staple items you can keep year-round to extend the life of produce, minimize waste, and help infuse quick meals with unique flavors.

Slow It Down to Soigné

I'm a visual learner, and my sensory experiences are led first with my eyes. In fact, I have horrible hearing, often having to read lips, so I've always relied far more on my palate and sight to compensate. Even with food and cooking, how ingredients or a composition look is a critical driving force behind my creative process and how I enjoy a dish. Flavor and texture must hit the right notes, but even if something is powerfully delicious but doesn't look right, it won't sit well with me. I even judge wine bottles by the label as much as the grape varietals and winemaking craft (which I know is sacrilegious to most sommeliers). To me, it's a component of the whole that must be integrated holistically.

Despite the high speed and chaos of service, there's always space to slow down and share a *chef's kiss.* Restaurants always put in that extra touch, that spark of magic that illuminates the entire symphony. A complimentary amuse-bouche to start off a meal, a vibrant, surprising garnish that jumps from a dish, a tableside preparation that brings you into the brigade, or a plate so beautiful you have to look underneath it to find out who made it. There's a theater and interactivity that glides through the performance, an art that often might feel just out of reach, exactly why you choose to eat out: to be entertained, to be nourished, to be taken care of.

"Soigné" is the back-of-house code word for when a restaurant wants to signal that a guest is special and often treated on the house. The term comes from the French verb *soigner*, which means "to take care of." Simple choices can be that cherry on top in the everyday. Taking that extra moment to arrange the plating in a particular order. Selecting ceramics that complement the color, texture, and shape of the vegetable atop. Serving something large-format family-style for the entire table to enjoy together. Layering in loads of dips and sauces for a choose-your-own-adventure course. Preparing a dish smaller as a delightful canapé for yourself or for guests as they arrive. Activating change requires a spark to inspire others. The imagery weaved throughout, anecdotes about the development of the recipes, and ideation around plating and serving are all meant to be points of reference for you to be the chef of your home restaurant. Nothing fancy, and no need to create an elaborate production. Just taking that extra moment of pause and sprinkling even a single little touch that emanates hospitality.

Zero Waste for Every Kitchen

"When you change the way you see things, the things you see change."
—Wayne Dyer

Now that we've talked about how we grow our food and choose what to source, consider that one-third of the world's food production is lost or wasted annually, often without much reason. For context, think about the full life cycle of food, from field all the way to shelf and then finally to disposal after it's purchased or eaten. About 10 percent of that journey's emissions come from wasted food. That equals the emissions of transit, and it's a piece of the system that we can affect every day in our kitchens.

Trash is all about perspective. There are a few small tactical shifts you can make at home in your kitchen and pantry that can support the environment. The first is that I challenge you to part ways with paper towels. Invest in some inexpensive but effective kitchen towels, and I promise you will be surprised by how quickly you can't imagine how you ever consumed so much paper before.

Next, collect all of the plastic elements in your home and consider swapping them for more sustainable materials. Research shows that we consume about a credit card's worth of plastic a week. Yes, you read that correctly—weekly. Plastic is cheap because we, through our tax dollars, subsidize that industry to make it cheap. But it's not inherently cheap, another myth of the Industrial Revolution. And we don't actually need it. There are beeswax alternatives to plastic wrap and plenty of options for functional and well-designed metal and glass storage containers. I keep a suite of jars of varying sizes for my preserves projects and to keep our pantry plastic free and well organized. All of my flours are in large glass jars with proper labels and dates for ease of visibility and tidiness. Invest in some blue masking tape and a Sharpie. Stasher silicone bags are

durable and reusable to replace your Ziploc bags, and I store stocks and frozen items (even leftover wine to cook with later on) in large silicone freezer molds to be pre-portioned when I'm ready to use them.

Finally, think about your cleaning products, another form of avoidable household toxins. There are so many effective and botanical brands out there, like Koala Eco, Grove Collaborative, and Blueland, that are safe for your family while getting the job done.

Let's Get Wasted and Talk Trash

My culinary philosophy is to use every part of the plant. Question what you might otherwise consider trash and ask yourself why. Understanding why something might be discarded is a first step in accepting ways to avoid throwing it away. Adopt a no-stalk-left-behind mentality and open yourself up to creative ways that scraps can serve some purpose. There are endless possibilities for innovation through upcycling.

Consider a few savvy substitutions, such as using shallots instead of red or yellow onion in most recipes. They're smaller, which means little to no waste and, let's be honest, when was the last time you used an entire raw onion in any dish? Plus, I actually think they taste better and sharper. Or, if a recipe calls for fennel, you might reserve the fennel frond and use as a garnish or as an herb; it tastes delicious with both sweet and savory dishes and looks pretty too. You can add the stalk to a soup for extra flavor or blend it into a sauce. Other dishes might use a whole vegetable as the serving vessel or turn to a mandoline to shave stalks into a dish. Peeling skin off produce is usually an unnecessary and wasteful step. If there's absolutely no way to incorporate a stem, leaf, or seed into a dish, it goes into stock or the compost.

Composting, the Basics

Compost is a natural fertilizer that improves soil's physical, chemical, and biological properties, made up of nothing more than broken-down plants and food waste and organic materials. Why compost? Landfills are the third largest driver of human-directed emissions. Throwing food into landfill releases 20 times the greenhouse gas emissions as composting. Yes, you read that right: *20 times*. Plus, when compost is not only collected but actually used, the carbon emissions are *negative*, as it's sequestering carbon as the compost feeds our soil.

No matter the size of your space or where you live, I promise you can compost. Sometimes it's hard to know where to begin. I hear that composting seems daunting, or gross, or there seems to be no space. There might always be a reason not to do something, but what if you can conjure up even one reason to try it? In our house we have a saying with our kids: Instead of "Why not?" let's ask "Why do?"

First things first. Get yourself a bin, with a lid—any bin will do. Put it on your countertop, proudly in the center for you and all to see. You can go simple from your hardware store or bougie with a Bamboozle. This is the very first step in a fresh journey to start collecting your food scraps. Make sure it's right where you cook so it's easy to use and visible as you prepare food throughout the week.

What you see, you can start to impact. Beginning first with just measuring what you throw away and could potentially have used will get your wheels turning. This also plays into the inherently competitive human spirit. Once you see the food waste accumulate weekly, you may start to compete with yourself and rise to the task of finding ways to throw out less and less over time.

What gets composted? Vegetable scraps should be collected separately and stored in your freezer for stock. Other food scraps, eggshells, and egg cartons as well as fibrous berry cartons from the market can be composted. Cardboard and even pizza boxes can be composted. Garden clippings such as grass and leaves as well as coffee grounds can be composted. Fibers, hair, and natural matter collected from the vacuum cleaners can also be composted.

Now finally, let's talk collection, which varies by location. A good place to begin is with community gardens and local farms, many of which offer drop-off opportunities. Many farmers' markets also offer collection points. Some cities do citywide curbside pickup or are testing municipal pilot programs (you can find information on public websites). For a smaller, private option, there are at-home countertop solutions emerging such as Lomi or Mill. If you're not familiar, Mill is a "white glove" composting option. The Mill odorless trash bin dries your organic waste and grinds it into nutrient-rich grounds that you can ship to be composted. For those looking to invest and create a truly circular system for your home garden, you can install a larger-scale composting system to feed your own soil. The US Environmental Protection Agency has a robust step-by-step program and process outlined for the public for the full adventure.

This is the final step in the regenerative cycle that we can have a hand in, literally. So, what are you waiting for?

We are now grounded. Grounded in the Earth that lies universally beneath our feet, that's home to all of our food, and grounded in these principles and practices to renourish our soil. When you're informed and

empowered to be part of the climate revolution, your cooking can be a formidable form of activism. The long-term solution to the impending climate crisis lies with our land, not in a lab, and each of us has so many choices each day to drive this change for our planet and for our health. Together we can regenerate for the next generation. So, in our home kitchens, let's be radicals, naturally.

To Store

Pickled Anything

Pickling has been a preservation technique since the beginning of time. It provides a hit of acid and salt and is a powerful way to make use of would-be food waste.

Ideas for what to pickle and keep at the ready in your refrigerator include carrots, cucumbers, strawberries, watermelon rind, beets, asparagus, mustard seeds, green onions, alliums of any kind (like leek, shallot, or red onion), and chilis. Any time you end up with leftover ingredients, you can pickle them to give them new life and bring a vibrant flavor to almost any savory dish.

The brine is equal parts vinegar and water, and must be enough to fully submerge the ingredients. Any vinegar will work as long as it's not aged or concentrated (like white vinegar or balsamic), and this is where you can have some fun. I love persimmon vinegar, and anything the company Tart Vinegar makes is special (especially their celery vinegar). You can mix vinegars together or stay pure with just one.

You will need a wide-mouth jar with a nonmetal lid (the acid can erode or rust a metal lid). Cut the vegetables or fruits into any shape and size you like so they fit snugly in the jar. Add any flavorings you desire. Options include strong fresh herbs (parsley, dill, rosemary, thyme), peeled garlic cloves, fresh ginger slices, whole spices (coriander seeds, peppercorns, mustard seeds), and chilis if you want it spicy.

To make the brine, in a small saucepan over high heat, combine equal parts water and vinegar with no more than 1 tablespoon kosher salt for every 2 cups (480 ml) liquid. Bring to a boil, stirring just until the salt dissolves. Pour the brine over the vegetables and aromatics, leaving ½ inch (1.25 cm) of room at the top. Allow to cool at room temperature with the lid off, then secure the lid and refrigerate for at least 48 hours before using. The pickles will keep for up to 3 months.

Preserves, Compotes, and Butters

The following techniques all help extend the life of any given produce. When I find something special at the market that has a short season, I often turn to these techniques so I can capture those unique flavors, transform them a bit through the preservation process, and use them in my cooking long after the fresh season has passed.

Preserved Lemons

Makes 1 quart (1 L)

A note of clarification: "preserved" here refers to a salt cure. (A sweet version of preserves can be found in the compote section.) The salt acts as a fermentation agent. Compared to fresh citrus, salt-cured lemons offer a nuanced umami and bright intensity of acid from the inside fruit and floral essence from the skin that you can't really put your finger on. All it takes is fruit, salt, and patience.

6 lemons
About ½ cup (80 g) kosher salt

Scrub the lemons and cut them into quarters lengthwise but not all the way through; keep the pieces still attached at the stem end. Pack them all into a glass jar, covering with salt as you add each piece. Seal the jar and put it in the refrigerator. Every 1 to 3 days, turn the jar upside down once or twice to reincorporate the liquid as it releases. The process takes at least 2 weeks. The flavor really starts to meld and become one between the rind and flesh—that's what you're looking for. Store in the refrigerator for up to 6 months.

PRO TIPS: I love doing this with Meyer or pink lemonade lemons.

I also utilize leftover lemons after juicing for this, though the entire batch should have some full fruit to get the benefit of the juice.

Another preserved fruit I use a lot in my cooking, often in salads or dressings, is umeboshi, a preserved Japanese ume plum, which is in season in June. The process is the same as above, sometimes with the addition of red shiso leaves for natural red coloring.

Berry Compote

Makes 1 quart (1 L)

We made this recipe through the restaurant as a custom product for the Glossier Berry Balm Dot Com launch as an in-store gift and to ship across the country. Their office was just around the corner, and we were embraced by their team from day one, as if we were their annex cafeteria. This is still one of our prouder creations, and it has a distinct balance of tart, sweet, and a pinch salty, perfect for porridge in the morning, with a cheese board, smothered on toast, or just by the spoonful. Try playing around with all kinds of berries you can find at the market. For the sweetener, coconut sugar, maple syrup, honey, and agave all work great; just be sure to adjust to taste.

1 pound (450 g) blueberries

3 ounces (85 g) mulberries

3 ounces (85 g) blackberries

2 tablespoons your choice
   of natural sweetener

¼ cup (60 ml) water

Grated zest and juice of 1 Meyer
   or regular lemon

Sea salt

Combine one-third of the blueberries, all of the mulberries and blackberries, the sweetener, and water in a medium saucepan. Bring to a boil over medium-high heat and cook, stirring from time to time, until the jam reaches 220°F (105°C), 15 to 20 minutes. Turn the heat down to low, fold in the rest of the blueberries, and continue cooking for another minute. The compote is meant to be chunky, with the taste of both cooked and fresh berries combined. Remove from the heat and mix in the lemon zest and juice and 2 pinches salt to brighten the flavors. Let cool to room temperature, then store in an airtight container in the refrigerator for up to 6 months.

PRO TIP: I use this exact method to make rhubarb compote, to have that beautiful tart floral fruit all year round.

Fruit Butter

Makes 1 quart (1 L)

A fruit butter is similar to a jam or jelly made without gelatin or preservatives. By design, it will extend the life of fresh fruit for months to come. Despite the name, there is no actual butter in this, it's just a metaphor for the silky-smooth spread that you will want to start putting on everything. I use it inside or on top of tea or breakfast cakes, spread on toast or baked goods, swirled into yogurt and oatmeal, or even just straight up by the spoonful. While it has a shorter shelf life than preserves, its brighter capture of the fruit brings memories of a different season year-round as the weather and light ebb and flow. It needs to be cooked low and slow. The apple provides a natural pectin that will bind and thicken the smooth butter.

Persimmons, pears, quinces, apricots, figs, cherries, peaches, and guavas are my favorite fruits to use for this. To prepare the fruit, remove all pits, seeds, and stems, then cut into medium-size pieces. Depending on the fruit, you may want to remove the skin (like guava), but for most it's OK to leave it intact.

6 cups pitted/seeded, cut-up fruit
  (see headnote)
1½ cups cored and cut-up apples
  (anything but green apples works)
¼ cup fresh lemon juice, orange juice,
  or apple cider vinegar
1¼ cups raw blue agave syrup
Optional seasonings: warm spices
  like ground nutmeg, cinnamon, or
  cardamom, or fresh botanicals
  like rose, elderflower, rosemary,
  or lavender

In a Dutch oven, combine the fruit, apples, and lemon juice. Bring to a simmer over medium-low heat and cook low and slow, regularly scraping the bottom and stirring with a spatula until the fruit is tender (a fork should go through easily), about 40 minutes.

Meanwhile, if you're using any optional seasonings, you can infuse them into the agave syrup. In a small saucepan, bring the agave to a boil and add the spices or botanicals. Remove from the heat. Let the flavors steep for 5 to 10 minutes, then strain.

Blend the fruit with an immersion blender or in a food processor until extremely smooth. Transfer the puree to a small saucepan, add the agave, and simmer over medium-low heat, stirring constantly, for about 30 minutes, until the butter starts to pull away from the bottom of the pan and is slightly thickened. Remove from the heat.

Let the mixture cool to room temperature, then transfer to airtight containers and store in the refrigerator for up to 6 months.

Lemon Kosho

Makes 2 cups (480 ml)

This is probably my favorite application of lacto-fermenting, in which simple ingredients come together for a layered, complex, heat-packed condiment. I have always had a deep love for yuzu kosho, a common ingredient in Japanese cooking. It's a paste made from fermenting yuzu zest and juice, green chilis, salt, and sugar. It is best to use Togarashi chilis here, but if you cannot find them, you can use serranos. I absolutely love yuzu and use it quite a lot in my cooking, but that said, I've adapted this recipe to incorporate lemons, which are standard for most home kitchens.

6 ounces (170 g) Preserved Lemons (page 30)
6 ounces (170 g) Togarashi chilis, stemmed and seeded
1 teaspoon honey

Combine all of the ingredients in the bowl of a food processor and blend until combined, about 1 minute, scraping down the sides intermittently. Transfer the mixture to an airtight container and store in a cool, dry place (about 60°F / 15°C is ideal). Check daily and taste for a sour tang; fermentation will be complete in about 2 weeks. Store in the refrigerator for up to 3 months.

On Vinegars

Vinegar is any liquid with 4 percent acetic acid. Vinegars are high in sugar, as they are made from fruits, honey, or starches like barley, potatoes, and rice. The ingredients are first fermented into an alcoholic product with the help of yeast, then oxidized into vinegar from there.

Here are some of my favorite vinegars to get your pantry started:

Camino red wine vinegar
Forvm chardonnay vinegar
Columela sherry vinegar
Tart golden vinegar
Figure Ate persimmon vinegar

Juice from Yumé Boshi organic umeboshi plums
Cabi sweet yuzu vinegar
Stone Hollow Farmstead apple cider vinegar
Rancho Gordo pineapple vinegar

Vegetable Stock
Makes 2 quarts (2 L)

For most savory recipes, using vegetable stock instead of water will immediately elevate your cooking and flavors. I often field questions about where to begin with stock, the dos and don'ts, and overall worry about getting it right. My aim is to demystify making stock so that it begins to circulate into your regular rotation.

The approximate ratio for a solid vegetable stock is 2 parts water, 1 part vegetables, and ½ part herbs and aromatics. The vegetables can be in any size or shape, because the stock will be strained in the end. I keep a bowl at the ready every time I prepare or cook something to collect all organic scraps, which I then separate into piles for compost and stock. Once you develop the routine, you'll realize how easy it is to upcycle trimmings into what will become a delicious, rounded-out base for so many recipes.

Focus on neutral, savory vegetables like alliums (leeks, onions, shallots), carrots, celery, fennel, mushrooms, and parsnips. You can include tomatoes, apple cores, and even eggshells, which add calcium and minerals to give strength and structure to the stock. Avoid vegetables that are too starchy like potatoes, artichokes, and turnips; water-based vegetables like cucumbers, zucchini, squash, and green beans that will break down and become too bitter when cooked so long; and intensely flavored vegetables like beets that would overpower the aroma and color. I also avoid brassicas like cabbage, broccoli, cauliflower, and kale as they can take over the balance of the flavor or get too bitter in a slow simmer.

Kombu or a Parmesan rind can be added as a thickening agent and for some umami, should you choose. Herbs that go best with a simple stock include parsley, thyme, bay leaves, marjoram, or rosemary. If you want a less-neutral stock, add aromatics such as black peppercorns, whole coriander seeds, or roughly chopped garlic, ginger, and lemongrass.

In a large stockpot, sauté about 5 cups of vegetable scraps in a bit of oil to get some golden color, then deglaze with a splash of wine for a few minutes until the alcohol evaporates (pro tip: you can freeze leftover wine in an ice tray to use for just this purpose). Cover with 10 cups (2.4 L) of water, or simply add the water to cover the raw vegetable scraps, which will render a subtler and clearer stock. Bring to a boil, then lower to a simmer. Add the herbs and any aromatics. Simmer for at least 2 hours, then strain out the solids and compost them. Cool the stock and store in airtight containers, labeled with the date, in your refrigerator (for up to 1 week) or freezer (for up to 3 months). I also like to pour some stock into ice molds and freeze, so that I can more easily use small amounts as needed.

Vegan Dashi

Makes 1 quart (1 L)

Japanese cooking has influenced my culinary perspective since I was a kid. As a professional cook, I have learned a tremendous amount from Naoko Takei, aka Mrs. Donabe, the owner of my favorite home goods store in Los Angeles, TOIRO Kitchen. Her graceful approach to one-pot cooking and the brilliant techniques she uses with her beautiful donabes are a revelation—that a simple broth can be elegant, built with a nuanced complexity of flavor, and made with ease even on a busy weeknight.

This dashi is something I turn to when I don't have vegetable stock in the freezer or the time to make a new one. It's also what I use when I want a subtle layered umami foundation on which to build a dish. It takes only about 30 minutes to make, and the ingredients are easy to keep on hand. Dashi is typically made with kombu seaweed and dried bonito flakes, but it can be made vegetarian by simmering kombu with dried porcini mushrooms. You can also use those ingredients to reinforce whatever vegetable stock you have on hand, following a ratio of 1 piece of kombu and 4 dried shiitakes per 1 quart (1 L) of liquid. You can make it hot (simmer together for 15 minutes) or steep in cold water for 2 hours, whatever is easier for you. I prefer to use the dashi right away, but you can store it in an airtight container for up to 1 week in the refrigerator and up to 2 weeks in the freezer.

1 quart (1 L) water or Vegetable Stock
   (page 35)
1 (12-inch/30-cm) piece
   dried kombu (kelp)
4 dried whole shiitake mushrooms

Bring the water to a simmer in a small saucepan over medium heat. Add the kombu and shiitakes and let simmer for about 15 minutes. Skim off any foam that rises to the top. Remove from the heat and let it steep for another 10 minutes, then strain out the kombu and shiitakes.

PRO TIP: Save the boiled kombu and use it in your next batch of Furikake (page 52) by dehydrating it, low and slow, in your oven.

Fermented Hot Sauce

Makes 1 cup (250 g)

Making a lacto-fermented hot sauce sounds complicated, but it's anything but. The gist is chopped chilis, salt, and a bit of water in a jar for 2 weeks. The result is a complex, layered heat that mellows the harsh spiciness of a raw pepper.

My favorite chilis include Fresno, Thai, serrano, Aleppo, ghost, Jimmy Nardello, lemon drop, and Scotch bonnet. The skins of the peppers contain the lactic acid bacteria needed for the fermentation. All vegetables and fruit can be lacto-fermented, as long as they are edible raw, making this a really fun and delicious use for leftovers.

1½ teaspoons kosher salt
1 cup (160 g) chilis, stemmed, seeded, and roughly chopped
1 clove garlic, peeled
Optional seasonings: salt or soy sauce, sugar or honey, vinegar

In a medium bowl, mix the salt with enough water to look like wet sand. Add the chilis and garlic and toss to combine. Transfer the mixture to an airtight container, cover, and store in a cool, dry place (about 60°F/15°C is ideal). Check daily and taste for a sour tang; fermentation will be complete in about 2 weeks. Once fully fermented, strain the chilis and garlic from the brine, transfer them to a blender, and blend until smooth. You can add a little of the brine if you want to thin out the sauce, but I prefer it thicker. If you like, you can season your fermented hot sauce with salt or soy sauce, some sugar or honey, and/or vinegar to round out the flavors. Store in an airtight bottle or jar in the refrigerator for up to 3 months.

PRO TIP: Once you've got the base down, you can start playing with fruit aromatics that are high in sugar (stone fruit, berries, melon) and whole spices (cumin seeds, cinnamon sticks, mustard seeds, bay leaves, coriander seeds), which would be put into the jar at the start of the recipe and fermented with the chilis.

Herb Oil

Makes about 1 cup (240 ml)

This is another great technique you can use with any fresh or almost expiring herbs, fig leaves, or scallions. The key with anything delicate is that you have to first blanch the greens in boiling water, then cool them in ice water to preserve their bright green color.

For a cleaner oil with no bits of herbs, you can strain it through a coffee filter inside a fine-mesh strainer. It will take longer for the oil to fully strain, but this will ensure it has no residue.

1 bunch chives, cut into 2-inch (5 cm) pieces (about 1 cup)
⅓ cup (4 g) fresh mint leaves
⅓ cup (4 g) fresh Thai basil leaves
¾ cup (180 ml) avocado oil

Bring a medium pot of water to a boil over high heat and prepare a medium bowl of ice water on the side. Add the chives, mint, and Thai basil to the boiling water and boil until the leaves turn bright green, about 30 seconds. Drain the herbs and transfer them to the ice bath. Once completely cooled, squeeze the herbs to remove as much water as possible.

Put the herbs in a blender and add the avocado oil. Blend on high speed until you get a fine puree, but be careful not to overblend, as that can heat the oil and leave it a less-vibrant green. Pour the contents of the blender through a fine-mesh strainer set over a bowl and let it drain undisturbed for a couple of hours, until the solids feel dry and most of the oil has passed through. Transfer to an airtight container and keep refrigerated for up to 1 week. Bring to room temperature before using.

PRO TIP: Stir the strained herb puree into any dressing or pesto for a fresh, herbal pop.

Chili Oil

Makes 1¾ cups (415 ml)

I have a tremendous collection of chili oils and crisps. Of course, Fly by Jing is one of my favorites, with its distinct and fiery Sichuan spice. We started to make chili oil at the restaurant to bring an amber warmth to one of our most popular dishes, the Sunset Grains, a hearty savory porridge made with Wild Hive Farm ancient grains and maitake mushrooms from Primordia Mushroom Farms in the Blue Mountain Ridge of Pennsylvania. I especially love the Basque Espelette pepper for its smoky notes. Thanks to a surprise care package in the mail, I connected with Boonville Barn Collective from Anderson Valley in Northern California, so I most often make this oil with their beautiful signature Piment d'Ville West Coast take on Espelette.

3 tablespoons + 1 teaspoon (30 g) ground Espelette pepper

3 tablespoons + 1 teaspoon (30 g) ground Aleppo pepper

2 teaspoons peeled and grated fresh ginger

2 teaspoons grated garlic

1½ cups (360 ml) avocado or toasted sesame oil

Combine all of the ingredients in a medium saucepan and bring to a simmer over medium-low heat, making sure not to let the raw ingredients burn. Lightly stir and simmer until fragrant, about 2 minutes, then remove from the heat and allow to cool completely. Store in an airtight container in the refrigerator for up to 6 months.

Mushroom Conserva

Makes ½ cup

Mushrooms can turn quickly if you don't use them. Preserving them in this conserva is an upcycling method to extend their use. While any mushroom will do, I particularly love to preserve special seasonal types of mushrooms like chanterelles and morels so I can cook with them year-round. The conserva is best served at room temperature. Pulse some to incorporate into sauces or spoon right over crusty toasted bread for a snack or appetizer.

¼ cup (60 ml) toasted sesame oil

1 cup (60 g) roughly chopped mixed
   fresh mushrooms

1 teaspoon kosher salt

½ cup (120 ml) soy sauce or tamari

3 tablespoons honey

1 tablespoon nutritional yeast

1 tablespoon peeled and grated
   fresh ginger

1 or 2 garlic cloves, grated

1 teaspoon finely chopped fresh thyme

1 teaspoon finely chopped
   fresh rosemary

1 tablespoon sherry vinegar

Heat the oil in a large skillet over medium heat until shimmering. Add the mushrooms and ½ teaspoon of the salt and cook the mushrooms until they are tender but not browned, about 10 minutes. Transfer the mushrooms to a large bowl and set aside.

Add the soy sauce, honey, nutritional yeast, ginger, garlic, thyme, rosemary, vinegar, and remaining 2½ teaspoons salt to the pan and return to the heat, stirring to combine until warm, about 5 minutes. Pour this marinade over the mushrooms in the bowl. Let fully cool, then transfer everything to an airtight container and refrigerate for up to 1 week.

Garlic Confit

Makes 1½ cups

This is a kitchen-sink technique to start incorporating flavored oils into your cooking (both hot and cold), which will add a layer to your flavors and can often bring vibrant colors to your plating. I use garlic confit in place of fresh garlic in any cooked recipe. Or spread on toast with a pinch of Maldon salt. It's great in dressings too.

½ cup (70 g) garlic cloves, peeled
1 cup (240 ml) avocado oil

Combine the garlic and avocado oil in a small saucepan and simmer over low heat until the garlic is tender, about 1 hour. Turn off the heat and let it rest until fully cooled. You can leave the garlic in the oil as an infusion or blend and strain it to have a composite oil with a rich color. Store in an airtight container in the refrigerator for up to 3 months.

PRO TIP: This technique works with anything fresh you want to preserve. The key ratio is 1 part vegetable to 2 parts oil. Some of my favorites are summer Sungold tomatoes and Jimmy Nardello chilis, so that I can have that bright flavor last into the colder winter months.

Crispy Fried Garlic

Makes ¼ cup

¼ cup (60 ml) avocado oil
¼ cup (35 g) garlic cloves, peeled
   and minced or thinly sliced
Kosher salt

Gently heat the avocado oil in a small frying pan or saucepan over medium heat. Add the garlic and fry until golden brown, taking care not to burn. Remove from the heat, drain, and season immediately with a pinch of salt. Once cool, store in a sealed container in your refrigerator or in a cool, dark place for up to a month.

Garlic Honey

Both honey and garlic are natural immunity boosters, and when combined they make the ultimate remedy for the common cold. But this homemade remedy doubles as a way to introduce a unique flavor profile to your cooking. The fermentation process here is easy; you just have to be patient, as the acidity of the honey naturally removes any worry of bacteria or toxins from the garlic.

Garlic honey gives a sweet funk with a little kick to any savory dish that needs a bit of sweetness. I often make a batch of this while I'm making garlic confit, ensuring I'm using all the beautiful garlic from the market. Just make sure you are using raw honey, as the natural bacteria and yeast help feed the fermentation process.

1 head garlic, cloves peeled
1 cup (240 ml) raw honey

Put the garlic cloves in a wide-mouth jar. Pour or spoon over the honey, ensuring the garlic is fully submerged. Cover the jar loosely with a lid and store in a cool, dark place. Every day, tighten the lid and turn the jar upside down and around for about a minute to recoat and move the contents around. Then loosen the jar lid and put it back in its storage spot. Keep this process going daily for at least 1 week but ideally for 1 month. The garlic honey will have fermented with a fully developed flavor and will be thinned out in consistency, as the liquid from the garlic is released over time.

Store in a cool, dry place for up to a year. The garlic flavor will mellow over time.

Citrus Salt

Makes 1 cup (263 g)

Making citrus dust is a wonderful way to utilize the whole citrus, drying the peels and grinding into a fine powder. It will allow you to sprinkle that vibrant citrus flavor onto anything you desire. I especially like to make citrus salt with it.

The key is to isolate the peels, remove the pith, and dehydrate the peels completely. You can certainly buy a dehydrator (they work wonders), but I'm not really one for gadgets, and we truly have no extra space for appliances in our efficient kitchen. Instead, we do it in the oven.

To make citrus dust, place the citrus peels on a rack set over a rimmed baking sheet and dry in the oven at 135°F (60°C) until leathery, 8 to 14 hours. Remove from the oven but leave the oven on. Let the peels cool, then grind in a spice grinder. Sift the ground peels through a fine-mesh strainer and grind the larger pieces again to ensure it's all a very fine, consistent texture. Scatter the powder on a parchment-lined rimmed baking sheet and dehydrate for another 30 minutes. Remove from the oven and let cool at room temperature, then store in an airtight container in the refrigerator for up to 1 week.

121 g coconut sugar
75 g Maldon salt
15 g citrus dust (see headnote)
15 g ground Espelette pepper
8 g citric acid

Combine all of the ingredients in an airtight container and store in the pantry for up to 1 month.

PRO TIP: Look for citric acid in the baking section of well-stocked grocery stores, or order it online.

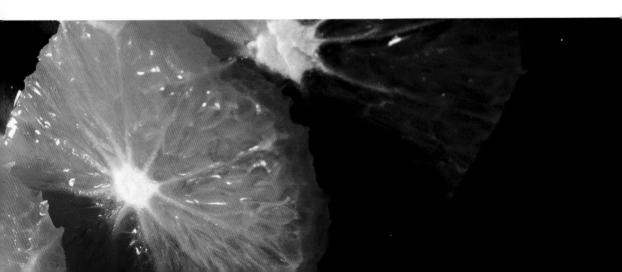

Pistachio Dukkah

Makes 1 cup

Dukkah is a Middle Eastern blend of herbs, nuts, and spices. Traditionally, the mixture includes hazelnuts, sesame seeds, ground cumin, coriander and fennel seeds, and salt. The balance of sweet and savory yields an exciting textural condiment for just about anything. We adapted a version of it at west~bourne when we had the restaurant, and I wanted to share our recipe.

This dukkah integrates high-protein flaxseed and hemp seeds for texture and health properties and centers the focus on one of the greatest superfoods, pistachios. New research has revealed that pistachios are a rare complete food, alongside avocados, high in nutrients like potassium and protein and in particular vitamin B6, which carries oxygen to red blood cells.

1 cup (140 g) shelled unsalted pistachios
½ cup (80 g) flaxseed
½ cup (85 g) hemp seeds
⅓ cup (45 g) white sesame seeds, toasted
¼ cup Shabazi blend (see Tip)
2 teaspoons kosher salt
1½ teaspoons garlic powder

In a small skillet, toast the pistachios over medium heat for about 5 minutes, until golden, toasty, and aromatic. Pulse them in a food processor until they are fine enough to sprinkle but not ground (they should not release their oils). Transfer to a medium bowl, add the remaining ingredients, and stir until well incorporated. Store in an airtight container at room temperature for up to 1 month.

PRO TIP: One of my favorite spice craftsmen, and a favorite among many chefs, is Lior Lev Sercarz of La Boîte in New York. His custom blends can be purchased online and are worth the investment for a completely distinctive seasoning for your everyday cooking. Lior's Shabazi spice mix is a classic Yemenite condiment that includes cilantro, dried green chilis, garlic, and lemon for a fresh hit of heat. If you don't have any on hand, you can substitute dried zhoug.

Crispy Bulgur Wheat

Makes 2 cups

Bulgur is a descendant of an ancient durum wheat grain, unchanged and dating back to the dawn of agriculture in the Fertile Crescent of the Middle East over 10,000 years ago. Environmentally, ancient grains increase crop resilience, as they are more adaptable to climates and soils with less water required, given their long heritage. They are inherently and naturally pest resistant, and they contribute to healthier biodynamic soil. Ancient grains are also functional foods, with rich folate content, fiber, and minerals.

This recipe is an all-purpose crunch that I use to top salads, crudo dishes, even roasted vegetables. It can provide a great texture and bite on top of ice cream and tarts too. Try it with quinoa instead of bulgur if you want something gluten free.

1 cup (160 g) bulgur wheat
1½ cups (360 ml) water
1 tablespoon avocado oil,
   plus 1 cup (240 ml) for frying
1 tablespoon sea salt

Combine the bulgur, water, 1 tablespoon of the avocado oil, and the salt in a medium saucepan and bring to a boil over high heat. Turn the heat down to a simmer and cook for 10 to 12 minutes, until the bulgur is cooked but still toothy. Remove the pan from the heat, cover, and let sit for 10 minutes to allow the bulgur to continue absorbing all the liquid. It should be al dente, like pasta. Pour the bulgur out of the pan onto a rimmed baking sheet and let cool completely.

In a clean saucepan, heat the remaining 1 cup (240 ml) oil over medium-high heat until hot but not smoking. Add ½ cup (80 g) of the bulgur, being careful not to overcrowd the pan; there will be some bubbling and steaming when it first hits the oil. Lightly mix the bulgur while cooking to ensure it doesn't clump together. Once the bulgur is crispy and golden brown, 5 to 10 minutes, use a slotted spoon to transfer it from the oil to a kitchen towel to drain. Immediately season with additional salt to taste. Repeat to cook the remaining bulgur.

Let the crispy bulgur cool completely, then transfer to an airtight container and store in a cool, dry place for a few days.

Furikake

Makes about ½ cup (140 g)

Furikake is a ubiquitous table seasoning in Japan. It's a sweet and savory blend of toasted sesame seeds, seaweed, and spices meant to be sprinkled on just about anything for instant umami. Sesame seeds are packed with healthy fats, protein, B vitamins, minerals, fiber, antioxidants, and other beneficial plant compounds. Nori is essential to a healthy ocean ecosystem and rich in several potentially health-promoting bioactive compounds, including antioxidants, flavonoids, and phenolic compounds.

1¼ ounces (35 g) toasted nori seaweed
2 tablespoons white sesame seeds, toasted
4¼ teaspoons kosher salt
2 teaspoons ground Urfa pepper
1¾ teaspoons garlic powder
1¾ teaspoons ground cumin
1¾ teaspoons Citrus Salt (page 47) or dehydrated citrus peel

Pulse the nori in a blender until finely ground. Transfer to an airtight container, add the remaining ingredients, and store at room temperature for up to 1 month.

Toasted Rice Powder

Makes about 2 tablespoons

¼ cup (50 g) uncooked white rice

Toast the rice in a dry skillet over medium-high heat, moving it constantly, until a deep brown color develops. Don't stop at golden brown; for the best flavor, you want it dark brown but not burnt. After it has cooled, transfer the toasted rice to a coffee or spice grinder, or a mortar and pestle, and grind into a coarse powder. Leftovers can be stored in a spice jar at room temperature for about a week, but it's best to make only what you will need.

Cashew-Date Milk

Makes 1 quart (1 L)

My husband recently discovered a lactose allergy, so I have been experimenting with ways to bake without dairy. This nondairy base offers a creamy alternative to regular milk that's rich but not too sweet. We keep it on constant rotation in our kitchen, and it's even incredible with coffee in the morning.

2½ cups (600 ml) filtered cold water
1¼ cups (200 g) raw cashews
¾ cup (135 g) Medjool dates, pitted (about 10)
6 tablespoons (90 ml) honey
1 teaspoon vanilla extract
½ teaspoon kosher salt
¼ teaspoon ground cinnamon

Combine the water, cashews, and dates in a large bowl and let soak at room temperature for 8 to 12 hours.

Pour the soaked nuts and dates, along with the soaking liquid, into a saucepan and heat over medium heat until steaming.

Transfer the mixture to a blender and add the honey, vanilla, salt, and cinnamon. Blend on low, gradually increasing to high, until completely smooth, about 1 minute.

Transfer to an airtight container and store in the fridge for up to 1 week.

To Toast

Shiso Meyer Lemonade Spritz

Makes 4 servings

Slowing down is an important part of mindful cooking, taking just that extra bit of time for intention. I love the slow-down culture of the South, where enjoying a homemade drink on the porch is an activity unto itself. The indulgent pastime of sipping lemonade for pure enjoyment inspired this spiked version. It is meant to be served in a large batch and shared, low and slow.

6 ounces (180 ml) fresh lemon juice
   (ideally from Meyer lemons)
1 ounce (30 ml) maple syrup
4 shiso leaves, thinly sliced (see Tip)
Maldon salt
3 ounces (90 ml) vodka
Ice
3 cups (720 ml) plain seltzer or
   sparkling water

Make the lemonade base by combining the lemon juice, maple syrup, half of the shiso, and a pinch of salt in a small bowl. Add the vodka, then let sit on the counter for 5 minutes to marry the flavors. Fill a pitcher with ice, then strain the lemonade mixture over the ice. Divide the lemonade evenly between four glasses, then top each glass with 6 ounces (180 ml) of seltzer. Garnish with the remaining shiso.

PRO TIP: Shiso (also known as perilla) is a Japanese mint that has a hint of anise. Known for its medicinal qualities and lovely tender leaves, it's beautiful and easy to grow. It's often found in Japanese grocery stores and, when in season, at farmers' markets.

Avocado Margarita with Frozen Avocado Pits

Makes 2 servings

This drink was inspired by a trip to Mexico, where I learned that avocado pits can be frozen for up to a year and function well as ice cubes, as they hold the cold for a while as you sip your drink. I loved the whole-fruit nature of the process, and it was a surprising new way to upcycle something that we often discard. Of course you can use a regular ice cube instead, but it's a delightful surprise moment for your guests when they discover what's been chilling their drink all along. A savory margarita with a fuller body from the avocado, this is meant to be strong and vegetal.

1½ ounces (40 ml) tequila
1 ounce (30 ml) Cointreau
1 ounce (30 ml) fresh lime juice
½ ounce (15 ml) maple syrup
½ avocado, pitted and peeled
¼ teaspoon fine sea salt
2 frozen avocado pits

Combine the tequila, Cointreau, lime juice, maple syrup, avocado, and salt in a blender and blend until smooth. Place the frozen avocado pits in two highball glasses and pour over the margarita. Serve right away.

Blood and Blackberry

Makes 2 servings

When we were dating, my husband took a deep interest in cocktail making, and almost every Friday night he would craft something that we hadn't tried before (one of the perks of dating before the legal drinking age is that drinking was mostly new to us both). I'll never forget the day he made a Blood and Sand. I don't often like sweet things, but it had just the right smooth balance, with a touch of smoke and bitterness to round it out.

Blood and Sand is one of the few classic mixed drinks that includes Scotch whisky. It's named after Rudolph Valentino's 1922 bullfighter silent film titled *Blood and Sand* as told in the 1930 *Savoy Cocktail Book*. My version is liquor heavy and ditches the juice (the "sand").

3 ounces (90 ml) Scotch

2 ounces (60 ml) cherry liqueur

2 ounces (60 ml) cassis (blackcurrant liqueur)

1½ ounces (45 ml) sweet vermouth (I love Dolin)

Ice

2 pieces orange peel

4 Luxardo or Fabbri cherries, for garnish

Combine the Scotch, cherry liqueur, cassis, and vermouth in a large mixing glass. Fill to the top with ice and stir vigorously until chilled and a little frothy (mixing it should be a light workout). For each drink, strain into a double rocks glass over a large ice cube. Twist an orange peel over the top of each drink to express the oils, then discard the peel and add 2 cherries to each glass.

Fresh Melon Ume Seltzer

Makes 1 serving

This homemade seltzer combines a few ingredients common to Japanese cooking that I absolutely love and that are staples in my home. Sweet, sour, and umami, it hits all the notes with a fresh herb finish from the shiso and a surprising bit of texture from the hemp or chia seeds. It's packed with vitamins and minerals for gut health in addition to looking quite striking on your table or bar.

1¼ ounces (40 g) cantaloupe, finely chopped
1 ounce (30 ml) yuzu juice
2 shiso leaves (see Tip, page 56)
1 teaspoon umeboshi (sour plum) vinegar
1 teaspoon honey
Ice
6 ounces (180 ml) plain seltzer or sparkling water
Hemp seeds or chia seeds, for garnish

Combine the cantaloupe, yuzu juice, and 1 shiso leaf in a small bowl and muddle. Add the vinegar and honey and mix to combine. Fill a highball glass with ice. Remove the muddled shiso leaf and pour the juice mixture into the glass, then top with the seltzer. Garnish with the remaining shiso leaf and sprinkle a pinch of hemp seeds over the top.

Matcha Pop

Makes 1 serving

I love the versatility of matcha (green tea powder) and its incredible health benefits from the high chlorophyll and amino acid content. Matcha is a superfood that boosts brain and liver function with a balanced dose of natural caffeine. I often turn to this drink for an afternoon pick-me-up or when I am craving a soda of some sort.

1½ teaspoons matcha powder
1 tablespoon hot water
2 teaspoons honey
6 ounces (180 ml) plain seltzer or
    sparkling water
Ice

Whisk the matcha powder with the hot water in a small bowl until completely dissolved and a little foamy, then whisk in the honey until dissolved. Fill a glass with the seltzer, fill nearly to the top with ice, then pour the matcha mixture over the ice.

PRO TIP: There are three grades of matcha: ceremonial, culinary, and ingredient. Though it's more expensive, I much prefer the ceremonial grade, especially from the Uji region of Japan, in this cold preparation. I particularly like Kettl and Rocky's matcha because of the intention and quality of sourcing.

Avocado Oil–Washed Martini

Makes 1 serving

My favorite cocktail is a dirty martini. I started experimenting with oil washing when we first launched our west~bourne avocado oil, especially with the extra-virgin, which lends a gorgeous forest-green hue and provides a rich coating from the healthy fat of the oil. Using the oil in the garnish too gives a pop of color and vibrant flavor that artfully complements the juniper notes of the gin.

Avocado Oil–Washed Gin
10 ounces (300 ml) gin
1 tablespoon avocado oil
1 or 2 rosemary or other herb sprigs

Martini
2½ ounces (75 ml) avocado oil–
    washed gin
½ ounce (15 ml) blanc vermouth
Fine sea salt
Avocado oil, for garnish
2 or 3 pitted green olives
    (ideally Castelvetrano)

A day before you plan to make the martini, start the gin infusion: Combine the gin, avocado oil, and herbs in a large jar or other airtight container. Shake vigorously, then place in the freezer to infuse for 12 hours. This will cause the oil to solidify and separate.

To make the martini, knock a hole in the top of the frozen oil in the jar and strain the mixture into a clean jar. Measure 2½ ounces (75 ml) of the oil-washed gin into a martini glass and add the vermouth and a pinch of salt, then stir to combine. Add a few small drops of avocado oil on top to garnish and finish with the olives.

Yuzu Vesper

Makes 1 serving

It's no secret that I have a deep love affair with yuzu. The balance of sour, bitter, soft, and punchy with a can't-put-your-finger-on-it floral citrus note adds just the touch of complexity I look for, especially in a cocktail. Yes, you can use lemon, I guess, but why would you if you can find yuzu?

The Vesper was crafted by James Bond in Ian Fleming's first book, *Casino Royale*, in 1953, a tribute to Bond's love interest, Vesper Lynd. This version is an ode to my beloved fruit and also a nod to the Yuzu-Sansho Sour served at one of my favorite night spots, Bar Goto, on the Lower East Side of New York. I used to hide away there on endless dates with my now-husband, so taken with it that I even once left my purse there, only to realize it the next morning.

1½ ounces (45 ml) gin
1 ounce (30 ml) yuzu juice
½ ounce (15 ml) vodka
¼ ounce (7.5 ml) dry vermouth
Ice
1 piece yuzu or lemon peel

Combine the gin, yuzu juice, vodka, and vermouth in a shaker and fill with ice. Shake, strain into a coupe glass, and garnish with the peel.

Chamomile–Almond Milk Punch

Makes 1 serving

Traditional milk punch serves up cozy winter vibes, and is one of my mom's absolute favorite holiday toddies. Think eggless eggnog, with milk, brandy, rum or bourbon, sugar, and vanilla served cold and with grated nutmeg on top. I wanted to lighten it up a bit for health qualities and make it a bit more seasonal year-round. This is equally satiating made without spirits, and I also love it blended with ice. Cashew milk can be used instead of almond if you have some on hand. Chamomile blossoms can be found with the edible flowers at well-stocked markets, or you can grow your own.

5 ounces (150 ml) almond milk,
   homemade (recipe follows)
   or store bought
1 ounce (30 ml) brandy
1 ounce (30 ml) dark rum or bourbon
1 teaspoon pourable honey
   or agave syrup
1 tablespoon fresh chamomile
   blossoms, plus more for garnish
2 drops vanilla extract
Ice

Combine the almond milk, brandy, rum, honey, chamomile, and vanilla in a cocktail shaker, fill with ice, and shake. Let the drink sit in the shaker for 5 minutes so the chamomile infuses. Fill a glass with ice, preferably crushed, and strain the drink over the ice. Garnish with additional chamomile flowers.

PRO TIP: You can infuse the honey hours before or even the night before if you want a stronger chamomile flavor. Just add the flowers to the honey and let it sit at room temperature.

Homemade Date-Sweetened Almond Milk

Makes about 6 cups (1.5 L)

If you're feeling fancy, try making your own almond milk at home.

1½ cups (210 g) almonds,
    raw and skinless
5½ cups (1.3 L) filtered cold water
4 pitted dates
½ teaspoon vanilla paste or seeds
    from 1 vanilla bean
Fine sea salt

Put the almonds in a bowl, cover with water by about 1 inch (2.5 cm), and soak overnight on your counter or for up to 2 days in your refrigerator. Drain the almonds and combine with the filtered water in a blender; blend on high for 2 minutes until smooth. Strain the milk, reserving the solids for another use (see Tip), then return the milk to the blender.

Put the dates in a small bowl, pour in just enough hot water to cover, and soak for at least 10 minutes, until softened. Add the dates and their soaking water, the vanilla, and a pinch of salt to the blender with the almond milk and blend on high for 30 seconds. Chill for at least 2 hours before enjoying. It will keep for up to a week in the refrigerator.

PRO TIP: The nut solids are delicious in overnight oats, or they can be composted.

To Begin

Leftovers Tempura with Ginger-Scallion Ponzu Sauce

Makes 2 servings

Tempura is an art form in Japanese cuisine, though interestingly enough, its origins trace back to Portugal. The dish was brought to Japan through trade relations in the sixteenth century, and the word *tempura* in Portuguese translates to "seasoning." During my last visit to Japan, one of our best meals was lunch at the notable Kondo, where they offer a singular dedication to the tempura craft, and it grew my obsession with perfecting the practice. I often make this with leftovers to give them a second life for a meal with my family, so save your scraps and give it a whirl. I like to serve this with two sauces for a contrast of flavors: a traditional ponzu-based one and some of my mushroom conserva tartar.

If you have the time, it's best to chill all of the dry ingredients in the freezer, which will produce the best crust. Some tempura batters call for ice water, but I prefer carbonation, which reduces the liquid amount and enhances the crispness of the coating (you can even use beer instead of seltzer).

Ginger-Scallion Ponzu Sauce

¼ cup (60 ml) toasted sesame oil

¼ cup (60 ml) ponzu sauce

1 tablespoon mirin

1 tablespoon peeled and grated
fresh ginger

1 tablespoon finely chopped scallions

Tempura

1 cup (170 g) cut raw vegetables,
roughly the same size

1 cup (115 g) chilled tempura flour,
plus more for coating

1¼ cups (300 ml) plain seltzer water
or club soda, chilled

1 large egg, chilled

1 teaspoon kosher salt

2 cups (480 ml) avocado oil

To Serve

Sea salt

Togarashi seasoning (optional)

½ cup (125 g) Mushroom Conserva
Tartar Sauce (page 85)

Combine all of the ponzu sauce ingredients in a small bowl and set aside while you prepare the tempura.

Dry the vegetables thoroughly, then lightly coat them in some of the tempura flour. The idea is to cover the vegetables in a light layer, not too thick. It helps the batter stick and ensures the vegetables are dry all around so they don't steam.

Loosely mix the remaining flour, the seltzer, egg, and salt in a medium bowl, ideally with chopsticks, but a fork will work too. You do not want the batter to be smooth; lumps are essential and it's key not to overmix.

Heat the oil to 350°F (180°C) in a deep frying pan; there should be ripples along the top of the oil, but it should not be smoking. To discern if it's ready, you can add a test piece of batter—it should bubble right away when placed into the hot oil. Once your oil is ready, quickly dip each piece of vegetable into the batter, allowing the excess to drip off before carefully transferring it to the frying oil. Fry the battered vegetables in batches so the pan is not crowded, which would cool the oil too much to render a crispy crust. Rotate the vegetables as they are frying so they turn evenly golden on all sides, then remove the pieces with a slotted spoon and place on a wire rack to drain. Season with salt while hot (and togarashi if you want a spice kick) and serve immediately with the ponzu sauce and tartar sauce for dipping.

PRO TIP: If you can't find tempura flour, you can substitute cake flour or rice flour.

Chili Egg

Makes 1 serving

There's nothing more decadent, and deceptively simple, than a jammy 6-minute egg. I love doing this as a one-bite canapé to kick off a dinner or family style for brunch. It's often a chef's snack for myself while I'm cooking, too. Try it with homemade Chili Oil (page 41) or Fly by Jing Chengdu Crunch, which is a particular favorite when I want a Sichuan kick.

1 teaspoon baking soda
1 large egg
Sea salt and freshly ground
   black pepper
Chili oil, homemade (page 41)
   or store-bought

Bring a small pot of water to a medium boil and add the baking soda, which will help release the shell from the egg. Gently place the egg in the boiling water and boil for 6½ minutes (if you prefer a slightly more set yolk, you can cook the egg for up to 8 minutes total).

While it's cooking, fill a medium bowl with ice and cold water. Transfer the cooked egg to the ice bath and let it cool completely. Peel the egg and slice in half lengthwise and place on your plate. Season with a pinch of salt and pepper and drizzle over the chili oil (don't be shy).

Garden Focaccia

Makes 1 18-by-13-inch (45 × 33 cm) focaccia

From the time I lived in Rome when I studied abroad in college, I've been inspired by the art of focaccia. It's a more forgiving entry point for those wanting to dip their toe into the craft of breadmaking. I developed this recipe as an interactive component to a lunch gathering, where each guest made their own. I baked the breads off while my guests ate their meal, and they could take the bread home as their own edible, self-created takeaway gift. It's a wonderful party trick and lets your friends get their hands dirty as part of the meal. Plus, focaccia is an ideal vessel to upcycle bits and pieces you might have left over into varying flavors, colors, and textures atop the bread. Have fun, use your scraps, and channel your inner artist—experiment and get messy!

7 cups + 7 tablespoons (930 g) all-purpose flour

3½ cups (840 ml) water

¾ cup (100 g) potato flour

1 tablespoon active dry yeast

¾ cup + 1½ tablespoons (202 ml) avocado oil, plus more for brushing the pan and finishing

3 tablespoons + 2 teaspoons (35 g) kosher salt

Toppings: herbs, edible flowers, zucchini blossoms, and/or thinly sliced vegetable scraps, plus more for garnish (optional)

Maldon salt

This recipe uses a poolish, which is essentially a young starter or pre-ferment, to help jump-start the fermentation of the bread in a short amount of time. The night before you want to bake your focaccia, combine 2½ cups plus 2 heaping tablespoons (330 g) of the all-purpose flour and 1 cup plus 7 tablespoons (341 g) of the water in a large bowl and stir until no dry spots of flour remain. Loosely cover with a towel and let sit at room temperature for at least 10 and up to 12 hours to ferment. It should be roughly doubled in size, with visible air bubbles.

The next day, add the remaining 2 cups plus 1 tablespoon water, remaining 4¾ cups plus 1 tablespoon (600 g) all-purpose flour, the potato flour, and yeast to the bowl of a stand mixer fitted with the dough hook. (The total amount of dough might be too much for a smaller mixer, so, if needed, you can mix the dough in two batches.) Mix on low speed until just mixed. Add the poolish and continue to mix on low until incorporated. Increase the speed to medium and knead the mixture for roughly 8 to 10 minutes. Turn off the mixer,

remove the bowl, and cover it loosely with a towel. Allow the dough to proof for 1½ to 3 hours at room temperature (or overnight in your refrigerator), until doubled in size.

Lightly punch down the dough and return the bowl to the mixer, still fitted with the dough hook. Add the oil in small increments, mixing between additions, then add the kosher salt. Knead on medium-low speed until just combined, about 5 minutes.

Brush a large rimmed baking sheet with oil, making sure to coat all sides and corners. With lightly oiled hands, spread out the dough on the pan, poking with your fingers to create a dimpled look, starting from the middle and working your way outward to the edges. It's very important to spread the dough out evenly, since this is the last time the dough will be touched. Let it proof on the pan, lightly covered with a towel or parchment paper, until the dough rises to at least the lip of the pan, 1½ to 3 hours.

Preheat the oven to 350°F (180°C).

Once the dough is fully proofed, drizzle a couple more tablespoons of avocado oil on top of the bread, then top with herbs, edible flowers, and/or vegetables, and sprinkle with Maldon salt. Bake for 20 minutes or until the bread is a light golden brown. Rotate the pan and increase the oven temperature to 425°F (220°C). Bake for another 15 to 20 minutes, until the bread is deeply golden brown and sounds hollow when tapped. Remove from the oven and let cool for 5 minutes, then transfer the pan to a rack to continue cooling for another 15 minutes. If you like, you can garnish with some more fresh flowers or herbs. Slice and serve warm or store in an airtight container at room temperature for up to 1 week.

PRO TIP: Active dry yeast will keep for a few months in an airtight jar or container in your refrigerator, so you can always have some on hand when the breadmaking urge strikes.

Zucchini Fries

Makes 2 to 4 servings

This recipe is a wonderful vessel for leftover pieces of vegetables, and offers a healthier alternative to traditional fries, with an added crunch from the panko. You can also forgo the breadcrumbs and just use oat or rice flour, which will give the vegetables a thin, crispy skin. I love to serve this with sheep's milk yogurt or coconut yogurt topped with a tablespoon of salsa verde for an herbaceous dipping.

1 zucchini
½ cup (46 g) oat or rice flour
1 large egg
Sea salt and freshly ground
    black pepper
½ cup (25 g) panko or
    homemade breadcrumbs
½ cup (120 ml) avocado oil

Trim the stem end off the zucchini and cut it in half crosswise. Stand each zucchini half up and slice vertically into ¼-inch-thick pieces. Stack the pieces, flat side down, then slice into ¼-inch-thick fry-shaped batons. You should have about 18 zucchini fries in total.

Make a breading station by setting up three separate shallow bowls and a rimmed baking sheet with a rack on top. Put the flour in one bowl. In the next, whisk the egg and season with salt and pepper. Put the breadcrumbs in the third bowl. Working with one piece of zucchini at a time, first coat the zucchini in the flour, then the egg, and lastly the breadcrumbs. Place the breaded zucchini on one half of the rack until all the breading is complete and you are ready to fry.

In a cast-iron skillet or other heavy-bottomed skillet, heat the oil over medium-high heat until shimmering but not smoking. Gently fry the zucchini in batches, so as to avoid crowding the pan, flipping them over so all sides become golden brown, 3 to 5 minutes.

Once they are fully cooked, transfer the fries to the other half of the wire rack to drain. Immediately sprinkle with salt and serve hot.

PRO TIP: You can swap the zucchini for eggplant, sweet potato, or squash, as they share high nutrient density and similar creamy texture.

Corn Ribs with Queso Fresco and Espelette Pepper

Makes 4 to 6 servings

This recipe uses every part of the corn from husk to cob. The process takes hardly any time at all, and the dish can serve as an easy snack or side dish.

I often call for Espelette pepper, an ingredient common in French cooking. I fell in love with a chili powder called Piment d'Ville, made from Espelette peppers grown regeneratively in California by the female-run Boonville Barn Collective. I will never forget the day that I got a handwritten letter from Krissy Scommegna, co-owner and founder of the collective, sharing the progressive practices used on her land in Anderson Valley and their unique position as one of the few California farms growing chilis at scale with this level of quality and sustainability. It took just one taste for Piment d'Ville to become a staple in my kitchen and cooking.

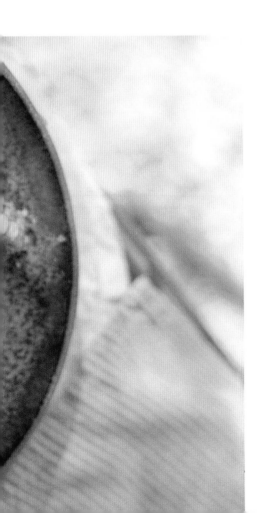

4 ears sweet corn

2 to 3 cups (480 to 720 ml) avocado oil or other neutral oil, for frying

Sea salt

Ground Espelette pepper

6 ounces (170 g) queso fresco, finely crumbled

¼ cup (10 g) fresh mint or shiso leaves, cut into chiffonade

Remove the corn from their husks. You can char the husks on your stovetop over an open flame for decorative plating, or compost them. Remove and compost the corn silk. Cut or manually break each ear of corn in half, then cut each half into quarters lengthwise, keeping the cores intact for each piece. The easiest way to do this is to stand each half cob up on your cutting board and cut down through the middle core, then lay each of those halves flat side down and cut in half again through the core to end up with four rib-looking pieces. You will have 32 ribs when finished.

In a deep pan, heat the oil over medium-high heat until it shimmers. Using tongs, carefully place the corn rib pieces, 4 to 6 at a time, into the oil to fry. They should immediately sizzle when they land; if not, your oil is not hot enough. If the ribs are not totally submerged in the oil, flip each piece as they cook. When the ribs are slightly browned, transfer them to a rack over a rimmed baking sheet to drain, then season immediately with salt and Espelette pepper. Repeat with the remaining ribs until all of them have been cooked and seasoned.

Place the corn ribs on a serving plate or bowl (lined with charred corn husks, if using), sprinkle the queso fresco over the top, and finish with the herbs.

PRO TIP: To chiffonade the herbs, roll a stack of the leaves and then slice thinly on a diagonal.

Vegetable Scrap Fritters

Makes 2 servings

This recipe is meant to be a super flexible base for any fritter, as a prime vessel for my creamy, umami-packed tartar sauce. Think of it as a vegetarian ode to the crab cake, which you can serve to start a meal or for an apéro hour, plus it can transform into a delicious sandwich. This can be made with almost any dense, hearty vegetable—fennel, carrot, turnip, sweet potato, broccoli, cauliflower, zucchini, or the like—and is a great way to incorporate vegetable scraps and leftovers. For additional protein, you can mix in some cooked lentils, if you like.

3½ ounces (100 g) vegetable of choice

1 shallot, finely chopped

1 teaspoon finely chopped
   toasted hazelnuts

1 teaspoon raw hemp seeds

1 teaspoon fennel seeds,
   lightly toasted and crushed

½ teaspoon ground cumin

½ teaspoon smoked paprika

¾ cup (45 g) panko or
   homemade breadcrumbs

⅓ cup (10 g) finely grated
   Parmigiano-Reggiano cheese

1 tablespoon finely chopped fresh basil

1 teaspoon citrus dust (see page 47)

1 teaspoon kosher salt

½ cup (43 g) chickpea flour

1 large egg, whisked

1 cup (240 ml) avocado oil, for frying

Sea salt

½ cup (120 ml) Mushroom Conserva
   Tartar Sauce (recipe follows)

Steam the vegetable until soft but still with a bite, being careful not to overcook or it will be mushy when processed. Pulse the vegetable using a food processor until chopped, or shred it on the large holes of a box grater, then drain or squeeze out any excess moisture. You should have about 2 cups vegetable.

Transfer the vegetable to a large bowl, add the shallot, hazelnuts, hemp seeds, fennel seeds, cumin, paprika, ¼ cup (15 g) of the panko, the cheese, basil, citrus dust, and kosher salt, and mix well. Shape the mixture into small patties and set aside on a plate.

Make a breading station by filling three separate shallow dishes: one with the chickpea flour, one with the egg, and one with the remaining ½ cup (30 g) panko. Working with one patty at a time, coat with the flour and dust off the excess, then dip into the egg, and finally cover in panko. Bread all of the fritters and place them on a plate until ready to fry.

In a wide, shallow skillet, heat the oil over medium-high heat until shimmering but not smoking. Gently drop the fritters into the oil, one at a time, working in batches as necessary so as to not crowd the pan. Fry until golden brown on the bottom, 3 to 5 minutes, then flip and fry on the other side. Drain on a wire rack over a rimmed baking sheet or on a kitchen towel. Immediately season with sea salt while still hot. Repeat until all the fritters are cooked, then serve with the tartar sauce.

Mushroom Conserva Tartar Sauce
Makes ¾ cup (180 g)

Use what you have on hand for this versatile recipe. For the labneh, swap in any plain yogurt you prefer. For the pickles, use anything briny you have on hand, such as pickled onions or cucumbers, olives, or capers. Tarragon, dill, or chives work well for the herbs. The idea is to give a mix of tart from citrus, brine and crunch from pickles or something like it, and vibrance from the herbs, all to bring out the umami of the mushroom conserva.

½ cup (120 ml) labneh
2 tablespoons finely chopped pickles
2 tablespoons Mushroom Conserva
(page 42), drained and
finely chopped
1 tablespoon fresh lemon juice
1 teaspoon chopped fresh herbs

Combine all of the ingredients in a small bowl and mix well. Use immediately, or store in an airtight container in the refrigerator for up to 1 week.

Bouquet Lettuce Wraps

Makes 4 servings

I had the privilege of dining at Maison by Sota Atsumi in Paris, and it was one of the most inspiring meals I enjoyed in 2023. Just as we sat down, we were each gifted a meticulously subtle, deceptively simple vegetable bouquet. I am not sure a small bite of raw vegetables has ever been so powerfully flavorful. It brought each of us to silence. This beginning garden bite is inspired by that moment when my mind was blown, and the beauty is that you can use small scraps from your cooking in this as well.

4 lettuce leaves
1 carrot, cut into thin matchsticks
1 Persian cucumber, cut into
   thin matchsticks
¼ cup (58 g) thinly sliced fresh fruit
   (see Tip)
¼ cup (25 g) pickled red onion
   (see page 28)
4 to 8 fresh chives
Mirin Vinaigrette (page 133)
Crispy Bulgur Wheat (page 50),
   Crispy Fried Garlic (page 45), or
   toasted white sesame seeds

Fill each lettuce leaf with a mix of the carrot, cucumber, and fruit. Top with the pickled onion, then, using 1 or 2 of the chives, carefully tie each lettuce cup closed. Place on a serving platter next to a small bowl of the vinaigrette. Sprinkle the bulgur, garlic, or sesame seeds over the top and serve.

PRO TIP: I prefer stone fruit here, but you can use apples, pears, or even oranges.

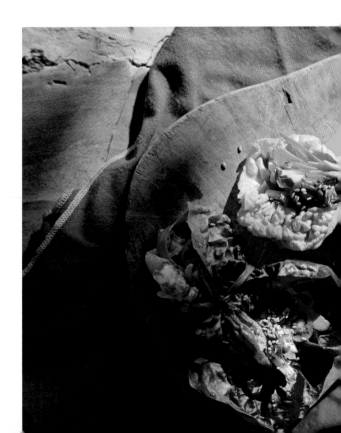

Popped Salted Sorghum

Makes 2 cups (260 g)

Making cover crops part of your cooking is easy once you get used to it. An interesting ingredient to work with is sorghum. Sorghum is a cereal grain that has high water-use efficiency, is resilient to hot climates, and is adaptable to climate fluctuations. Here, we pop the dried kernels for a more sustainable popcorn-inspired snack with a delicate, sweet kernel.

1 tablespoon avocado oil
2 tablespoons dried sorghum kernels
Sea salt or Himalayan pink salt

Heat the oil in a large, deep pot or Dutch oven over medium-high heat until shimmering. Add the sorghum, cover, and reduce the heat to medium-low. Keep moving and shaking the pot as the kernels pop to prevent burning. Once the popping slows down, turn off the heat and continue shaking the pot with the lid on for 5 to 10 more seconds in case there are some late poppers. Carefully remove the lid and transfer the popped sorghum to a bowl. Season with salt as desired.

PRO TIP: As a flour, sorghum is smooth and mild in flavor, forming an ideal base, akin to a wheat flour, for naturally gluten-free baking. It binds moisture in doughs and batters and can promote carbon dioxide bubbles when used to make bread. It should be combined with another flour and kept to less than 30 percent of the total flour in baking, or it can impart a sour flavor.

Sea Salt–Vinegar Ginkgo Nuts

Makes 2 to 4 servings

Think of this as a more nutritious and interesting bar snack—a discoverable option in place of peanuts, popcorn, or dare I say, corn nuts. Ginkgo nuts are the fruit of the ginkgo tree, a highly adaptable, happily urban plant originally from Japan but also common across Brooklyn. Super umami and carrying medicinal qualities, the nut is encased in a hard shell. You can find them shelled and canned or vacuum sealed at Asian grocery stores, but I recommend getting your own in-shell nuts, which you must cook before eating the inner nut (you can find them at Japanese markets or online). It is important to note that raw ginkgo nuts are toxic, so be sure they are properly prepared before eating!

The ginkgo nuts will be pan fried in batches, so be careful not to crowd the pan. The nuts will also tend to pop once the shell breaks during cooking, so it's important to use a splatter screen or even a baking sheet over the pan while cooking.

8 tablespoons (120 ml) avocado oil
2 cups (300 g) raw, shell-on ginkgo nuts
Sea salt
¼ cup (60 ml) white or
    champagne vinegar

Heat 2 tablespoons of the avocado oil and ½ cup (75 g) of the raw nuts in a cast-iron skillet over medium-high heat. Season with a heavy pinch of salt. Cover the pan and gently shake it to keep the nuts moving around, similar to how you would cook popcorn, and cook until the shells split and the inner nut is a soft green. Once the popping has stopped and most of the nuts look cracked, transfer them to a bowl. Repeat this process three more times, until all the ginkgo nuts are cooked. If any nuts do not open during cooking, you can crack them with your hands or a heavy utensil, but be careful since they will still be hot.

Once you have all the cooked nuts in the bowl, season with the vinegar and more salt to taste. Serve warm.

Vibrational Crispy Chickpeas

Makes 2 to 4 servings

Chickpeas are indeed a superfood. One of the earliest cultivated crops in history, they are high in both fiber and protein. This dish started one day when I had leftover chickpeas from a salad. I crisped and spiced them up into a vibrant, tempting bar snack or accompaniment to a drink or as part of a picnic. These chickpeas can also be used instead of croutons for crunch in a salad. Though I must admit, it's a struggle to wait to devour them.

1½ cups (270 g) dried chickpeas or
    1 (15.5-ounce/440-g) can chickpeas
½ to 1 cup (120 to 240 ml) avocado oil
1 tablespoon nutritional yeast
1 teaspoon ground turmeric
½ teaspoon ground white pepper
Sea salt

If your chickpeas are dried, put them in a bowl, cover them with 1 inch (2.5 cm) of water, and leave them out on the counter overnight. Then drain, rinse, and dry them well. If your chickpeas are canned, simply drain and rinse them, then dry completely. You want to be sure they are not wet so the oil doesn't splatter while frying.

Set up a place to drain the chickpeas after frying, either a rimmed baking sheet with a rack on top or a kitchen towel. In a cast-iron skillet or other large, deep pan, pour in enough avocado oil to cover the chickpeas at least halfway up. Heat the oil over medium-high heat until hot but not smoking. Fry the chickpeas in small batches, without crowding the pan. Once the chickpeas are crispy and golden brown, about 3 minutes, use a slotted spoon to transfer them to the rack or towel. Once all the chickpeas are cooked, put them in a medium bowl and toss gently with the nutritional yeast, turmeric, and white pepper. Taste and adjust the spices and season with salt. Serve immediately. You can store these in an airtight jar for a day or two, but they are best fresh.

Crispy Za'atar Socca with Shallot Labneh

Makes two 5-inch (12.5 cm) pancakes

Socca, or farinata, is an unleavened pancake or flatbread made of chickpea flour, with origins in the northern Mediterranean coast in Liguria, between the French and Italian border. Chickpeas are gluten free and a rich source of protein, and they also carry a very small carbon footprint given that the plant can fix nitrogen for itself and thrives in diverse soil conditions. Chickpeas are also a break crop, providing a natural repellent against weeds and disease and protecting other crops in the ecosystem on a farm.

If you want to increase the recipe, the ratio is equal parts chickpea flour and water with one-quarter part avocado oil for the batter. For an added layer of flavor, you can add other spices, such as cayenne, ground cumin, or curry powder (½ to 1 teaspoon total), to the batter.

½ cup (43 g) chickpea flour
½ cup (120 ml) water
½ teaspoon kosher salt
½ teaspoon freshly ground
  black pepper
3 tablespoons extra-virgin avocado oil
Grated zest and juice of
  1 Meyer lemon
½ cup (120 ml) Shallot Labneh
  (recipe follows)
Za'atar
Leaves from 1 mint sprig

Combine the chickpea flour, water, salt, pepper, and 2 tablespoons of the oil in a medium bowl and whisk until smooth. Cover the bowl with a towel and let it sit at room temperature for at least 1 hour or as long as 12 hours (let it rest as long as you can within that time frame to allow the flour to fully hydrate).

After the batter has rested, heat the remaining 1 tablespoon oil in an 8-inch (20 cm) cast-iron skillet over medium-high heat. Pour half of the socca batter in the center of the hot skillet (like you're making a pancake) and let it spread out. It should create about a 6-inch (15 cm) pancake covering almost the entire bottom of the skillet. Let it cook until golden brown and crispy on the bottom, about 2 minutes (increase the heat if necessary, but be careful not to burn the pancake or oil), then use a spatula to flip to brown and crisp the other side for about 1 minute. When it's done cooking on both sides, transfer the pancake to a plate. Repeat with the remaining batter.

Stir the lemon zest and juice into the shallot labneh and spoon it over the socca, then sprinkle za'atar on top to finish. Garnish with the mint leaves.

PRO TIPS: Top with a fried egg for a warming savory breakfast or make into small silver-dollar pancakes as a delicious gluten-free base for canapés. I also love making a few of these and surrounding the table with an array of condiments such as tomato chutney, seedy tahini, zhoug, or Walnut Bagna Cauda (page 136) for an interactive first course, allowing guests to choose their own adventure among the dips.

Shallot Labneh

Makes 1 cup (240 ml)

This was a staple condiment at west~bourne restaurant. It's loosely based on compound butter, where additional finely chopped ingredients and spices are incorporated to add flavor to a rich, creamy base. You can easily swap the labneh for Greek yogurt, sheep's milk yogurt, White Moustache yogurt (if you can find it), crème fraîche, or even a thick neutral coconut yogurt—anything with a thick, smooth texture and that is rich and tangy and doesn't have any added sweetener. The shallot layers in a crunchy texture and also gives a savory hint to the tang of the yogurt. This is meant to be a core recipe that you can expand on and experiment with. Play around with flavors and combinations such as added spices, grated garlic, or herbs as you wish.

I use this with just about everything savory—on breakfast tacos or under roasted vegetables—and I even turn it into dressing. It makes a great dip for crudités, too.

1 cup (225 g) labneh or Greek yogurt
2 tablespoons minced shallot
4 teaspoons extra-virgin avocado oil
Sea salt and freshly ground
   black pepper

Put the labneh in a medium bowl, fold in the shallot and oil, and season with salt and pepper to taste. Store in an airtight container in the refrigerator for up to a few days.

PRO TIP: I recommend Karoun brand labneh.

Charred Jimmy Nardello Peppers with Crème Fraîche and Herb Salad

Makes 2 to 4 servings

If you haven't yet had a Jimmy Nardello pepper, it's a revelation. It's like a love child/hybrid of all the best peppers ever grown in the world: spicy, sweet, earthy, and tangy all at the same time. I became so obsessed that we planted them plentifully in our home garden, and they barely make it off the vine before they are devoured by the basketful.

Peppers have a deep history across the world, grown in many regions as an affordable alternative to black pepper to add spice to dishes. In 1887, Jimmy Nardello's mother brought these pepper seeds with her from her hometown in the Basilicata region of southern Italy and started growing them in her humble garden in Connecticut. Jimmy, the fourth of eleven children, donated his mother's seeds to Seed Savers Exchange before he died in 1987, and later, Jimmy Nardello peppers were given a spot on the Slow Food Ark. Often known as the perfect sweet Italian frying pepper because they melt into a luscious, creamy, soft delight when fried, they have become cult favorites among farmers and chefs.

I love them charred and softened with vinegar, dipped while hot into cold, creamy crème fraîche, and topped with a bright herby salad. For the herbs, you can use tarragon, mint, basil, parsley, Thai basil, shiso, or cilantro.

8 ounces (230 g) Jimmy Nardello peppers

½ cup (120 ml) sherry vinegar

Sea salt and freshly ground black pepper

1 cup (12 g) fresh herbs

Grated zest and juice of 1 lemon

5 ounces (140 g) crème fraîche

Crispy Fried Garlic (page 45), to garnish (optional)

Maldon salt

2 tablespoons extra-virgin avocado oil, Garlic Confit oil (page 45), or Nardello oil

Fire up your grill or, if you have a gas stove, turn the burner on your stovetop up to high. In batches, place the peppers on the grill or directly on the stovetop flame. As they char and blister, use tongs to rotate them so they get charred all the way around, continuing until all the peppers are charred and softened throughout.

Transfer the peppers to a large bowl, immediately dress them with the vinegar, and season with sea salt and pepper. Cover the bowl with a kitchen towel and let the peppers marinate for about 5 minutes.

Meanwhile, prepare the herb salad by combining the herbs in a bowl with the lemon zest and juice, then season with sea salt. Let sit for a few minutes to allow the flavors to meld.

To plate, swirl the crème fraîche across the bottom of a serving platter. Arrange the peppers in a gentle mound in the center. Top the peppers with the herb salad and sprinkle with crispy garlic, if desired. Top with a pinch of Maldon salt and drizzle with the oil to finish.

PRO TIP: Crème fraîche is one of my top ten favorite condiments. I enjoy it sweet, savory, or by the spoonful. I am a huge fan of Bellwether Farms for the European-style richness, simplicity, and quality of ingredients in their products. That said, you can actually make crème fraîche at home. Simply combine 1 tablespoon buttermilk with 1 cup (240 ml) heavy cream in an airtight container and let it rest at room temperature for 12 hours. You can store the crème fraîche in the refrigerator for up to 2 weeks. The flavors will bend based on the kinds of cream and buttermilk you use and how long you let it sit out.

Smashed Sweet Potatoes

Makes 6 to 8 servings

Sweet potatoes, especially Japanese sweet potatoes, are among my favorite produce and a powerhouse superfood: high in fiber and antioxidants, while packed with vitamins A, C, and B6 to promote heart health. This recipe was one of the most popular and beloved from my restaurant west~bourne, where we served it morning, noon, and night. I use a donabe with a steamer basket to steam the sweet potatoes.

6 sweet potatoes, about 1½ pounds (680 g), well scrubbed
½ cup (113 g) unsalted butter
½ cup (135 g) white miso
½ cup (20 g) fresh Thai basil leaves
½ cup (20 g) fresh mint leaves
½ cup (20 g) fresh cilantro leaves
2 tablespoons avocado oil, plus 1 to 2 cups (240 to 480 ml) for frying
1 tablespoon fresh lemon juice
Sea salt
2 ounces (55 g) crème fraîche
½ cup (60 g) Crispy Fried Garlic (page 45)

Pour an inch or two of water into a large pot, add a steamer rack, and bring to a boil. Place the sweet potatoes on the steamer rack, cover, and steam for 20 to 30 minutes, until tender. If the potatoes are large, they may take longer, so check them by poking with a cake tester or small knife; the tester should come out clean with no resistance. Remove the potatoes from the steamer and set aside until cool enough to handle.

To make the miso butter, preheat a small saucepan over medium-low heat, then add the butter. Swirl the pan occasionally to be sure the butter is cooking evenly. As the butter melts, it will begin to foam. Cook for 2 to 3 minutes, whisking as you go to ensure the solids do not stick to the bottom of the pan. The color will progress from lemony yellow to golden tan to, finally, a toasty brown. Once you smell a nutty aroma, take the pan off the heat and transfer the browned butter to a heatproof bowl to cool slightly, 5 to 10 minutes. Blend the brown butter and miso in a blender or small food processor until combined, then set aside in a large bowl.

Wash and spin dry the basil, mint, and cilantro, then toss the herbs together in a small bowl. Season with the 2 tablespoons avocado oil, half of the lemon juice, and salt to taste. Set aside.

Carefully tear the sweet potatoes into medium-size chunks with your hands, exposing the flesh and creating ridges that will allow for crispy edges. Pour enough of the remaining avocado oil into a cast-iron or other heavy-bottomed pan to a depth of about 1 inch (2.5 cm) and preheat it over medium heat. Once it's shimmering, fry the sweet potato chunks in batches until the edges are crispy. Transfer the cooked sweet potato chunks to a rack over a rimmed baking sheet to drain and immediately season with salt. Once all the sweet potatoes are finished, add them to the bowl with the brown butter miso. Gently fold the potatoes with a spoon so that they are evenly coated with the paste.

To serve, spread the crème fraîche over the bottom of a serving bowl or platter and pile the fried sweet potatoes on top. Drizzle the remaining lemon juice over the potatoes, and garnish with the herb salad and fried garlic.

Savory Galette

Makes 6 to 8 servings

A galette is a fancy French word for a free-form pie that requires no hardware (read: easy, just go with it). This dough is made from cover crops and is naturally gluten free. The idea is to build up the composition of the galette, almost the way you would put together a pizza. For the base, you can sprinkle low-moisture cheese or even pesto. This helps create a barrier between the vegetable and the crust to prevent the moisture from the vegetables from dissolving the crust, plus it adds a nice layer of fat, creaminess, and texture, and a slightly umami note. Thinly slice any vegetable you'd like, such as tomato, zucchini, potato, mushrooms, and so on. You can sprinkle all kinds of toppings on the folded edges of the galette, like toasted sesame seeds, ground cumin, nutritional yeast, za'atar, dukkah, or just salt and pepper.

This recipe can be easily converted to a sweet variation (see page 198).

⅓ cup (40 g) whole-grain buckwheat flour

¼ cup (27 g) whole-grain oat flour

¼ cup (27 g) sorghum flour

⅛ teaspoon sea salt

⅔ cup (160 ml) avocado oil

1 cup (240 ml) water

1 large egg white, beaten

3 tablespoons pesto or shredded cheese, such as Gruyère, Comté, Parmigiano-Reggiano, or cheddar

2 cups thinly sliced vegetables

Toasted sesame seeds, ground cumin, nutritional yeast, za'atar, dukkah, or freshly ground black pepper, for topping (optional)

Combine the buckwheat flour, oat flour, sorghum flour, and salt in a large bowl. Add the oil and stir until combined. Add the water, 1 tablespoon at a time, and gently mix to form the dough into a ball. Remove the dough from the bowl and place on a clean work surface. Form the dough into a disk, wrap with parchment paper, and refrigerate for at least 1 hour and up to 2 days (you can also freeze the dough in an airtight container for up to 3 months).

When ready to bake, preheat the oven to 350°F (180°C). Line a rimmed baking sheet with parchment paper.

Roll the dough disk between two sheets of parchment to roughly ¼ inch (6 mm) thick. Transfer the free-form shape to the prepared baking sheet. Brush the egg white across the whole surface of the dough (this will seal it and prevent the bottom from getting soggy), saving a little for when you're ready to close up the galette. Let the egg white dry, about 2 minutes, then spread the pesto or sprinkle the cheese across the dough, from the middle to the outside, leaving a 1-inch (2.5 cm) border to fold over the filling later. Neatly shingle the vegetables on top of the pesto or cheese. Use your fingers to fold the edges of the crust over the filling, leaving the center uncovered. Brush the crust with the remaining egg white and sprinkle on any spice or topping you might want to add.

Bake for 1 hour, until the edges start to brown and crisp. Remove from the oven and cool on a wire rack before serving.

PRO TIP: Be sure to cut the vegetables thinly so that they properly cook and brown.

Mushroom Larb Lettuce Cups

Makes 3 servings

This quick, go-to canapé or family-style starter packs heat and crunch. Larb is typically made of ground meat and lots of scallions and fresh herbs. It is dressed with lime juice, fish sauce, dried chili flakes, and toasted rice powder. Here, I'm doing this method with mushrooms to mirror the robust umami. Mushrooms and their fungi networks are critical to soil health, providing necessary nutrients and water to plants that are often removed by conventional tilling and pesticides. You can use any mushrooms of your choice, and a mix is lovely.

1-inch (2.5 cm) piece fresh ginger, peeled and grated

Juice of 2 limes

1 tablespoon soy sauce

1 teaspoon honey

4 Thai chilis, stemmed and thinly sliced

3 green cabbage or sturdy lettuce leaves

8 ounces (230 g) mushrooms

1 tablespoon avocado oil or Garlic Confit oil (page 45)

1 shallot, finely chopped

1 lemongrass stalk, outer leaves removed, finely chopped

1 garlic clove, grated

1 tablespoon Toasted Rice Powder (page 52)

Sea salt and freshly ground black pepper

2 mint sprigs, leaves picked

2 Thai basil sprigs, leaves picked

1 scallion, thinly sliced

In a small bowl, whisk half of the grated ginger with the lime juice, soy sauce, honey, and chilis. Set the dressing aside.

Prepare an ice bath in a large bowl. If you are using green cabbage, cut the head in half through the core, remove the limp outer leaves, and carefully peel off three perfect sturdy leaves. Place them in the ice bath to firm up until ready to eat. If using lettuce, peel off three large leaves and dunk them in the ice bath to firm up before drying with a towel. Let these sit while you prepare the filling.

Pulse the mushrooms in a food processor or chop until you have pea-size pieces. Heat the oil in a large skillet over medium-high heat until shimmering. Add the remaining ginger and the mushrooms, shallot, lemongrass, and garlic. Sauté until the mushrooms are cooked through and the mixture is well combined. Season with the toasted rice powder and salt and pepper to taste. Transfer the mixture to a towel-lined plate to drain.

If using cabbage, remove it from the ice bath and dry with a towel. Divide the mushroom filling among the three dry cabbage or lettuce leaves. Pour a little dressing over each larb cup, then garnish with the mint, Thai basil, and scallion. Serve, with the remaining dressing on the side.

Sweet or Savory
Toasty Seeded Granola

Made naturally gluten free using oats, a cherished cover crop, and intentionally crispy and golden toasted rather than blond, this granola is packed with superfoods like chia, flax, and tahini. A few small changes easily shift it from sweet to savory, the latter great alongside a cheese board or for an apéro hour as an evening snack.

Sweet Version
Makes 5 cups (700 g)

2½ cups (240 g) rolled oats

1 cup (120 g) sliced or slivered raw almonds or hazelnuts

½ cup (75 g) chia seeds

⅓ cup (52 g) flaxseed

2 teaspoons kosher salt

1½ teaspoons ground cinnamon

2½ tablespoons white miso

1½ tablespoons melted unrefined coconut oil, cooled

⅓ cup + 1 tablespoon (90 ml) maple syrup

5 tablespoons (75 ml) avocado oil

2 tablespoons honey

⅔ cup dried fruit (golden berries, tart cherries, etc.—something healthy and not overly sweet) or chocolate chips (optional)

Preheat the oven to 325°F (165°C). Line a rimmed baking sheet with parchment paper.

In a large bowl, combine the oats, nuts, chia seeds, flaxseed, salt, and cinnamon and stir well. In a separate bowl, mix together the miso and coconut oil, breaking the miso up with a silicone spatula. Whisk in the maple syrup, avocado oil, and honey. Add the wet mixture to the dry and mix to combine.

Spread the granola mixture evenly on the prepared baking sheet, then bake for 25 to 30 minutes, rotating the pan halfway through to ensure even cooking. The granola should be golden brown and toasty (not dark or mahogany brown). Though it may look a little wet when you remove it from the oven, be sure to let it cool completely before eating, as it will dry and get crunchy as it cools. The granola can be broken up or left in small chunks, then stored in an airtight container in a cool, dark place for up to 1 month. To preserve the freshness and crunch of the granola while stored, add the dried fruit or chocolate chips (if using) when you are ready to eat.

Savory Version
Makes 4¼ cups (515 g)

2½ cups (240 g) rolled oats

¾ cup (97 g) sliced or slivered raw
  almonds or hazelnuts

3 tablespoons (25 g) chia seeds

3 tablespoons (24 g) flaxseed

1½ teaspoons ground cumin

½ teaspoon kosher salt

5 tablespoons (50 g) Chili Oil
  (page 41)

¼ cup (61 g) tahini

2 tablespoons honey

1 tablespoon white miso

Preheat the oven to 325°F (165°C). Line a rimmed baking sheet with parchment paper. In a large bowl, combine the oats, nuts, chia seeds, flaxseed, cumin, and salt and stir well. In a separate bowl, whisk together the oil, tahini, honey, and miso until smooth. Add the wet mixture to the dry and mix to combine.

Spread the granola mixture evenly on the prepared baking sheet, then bake for 30 to 40 minutes, rotating the pan halfway through to ensure even cooking. The granola should be golden brown and toasty (not dark or mahogany brown) and nutty smelling. Though it may look a little wet when you remove it from the oven, be sure to let it cool completely before eating, as it will dry and get crunchy as it cools. The granola can be broken up or left in small chunks, then stored in an airtight container in a cool, dark place for up to 1 month.

Buckwheat Dutch Baby

Makes 4 to 6 servings

I had long revered buckwheat in soba noodles, but I only started to understand its role in baking when I first visited Blackberry Farm about fifteen years ago. Technically, it's not even a grain. Buckwheat is a seed harvested from a flowering plant that's a relative of rhubarb, which happens to be one of my favorite vegetables. Compared to grains, buckwheat has a distinct amino acid composition that supports heart and gut function.

The key to this Dutch baby is a hot pan and a loose, smooth batter. This recipe is naturally gluten free and intended to highlight buckwheat and oat, key cover crops that are as nourishing to our soul as to our soil.

¾ cup (70 g) oat flour

2 tablespoons buckwheat flour

2 packed tablespoons
dark brown sugar

1½ tablespoons brown rice flour

1 teaspoon kosher salt

½ teaspoon ground cardamom

¼ teaspoon baking powder

⅛ teaspoon baking soda

3 large eggs, at room temperature

7 tablespoons (104 ml) milk,
at room temperature

¼ cup (55 g) unsalted butter, at room
temperature and cut into pieces

Preheat the oven to 425°F (220°C) with a 9- to 10-inch (23 to 25 cm) cast-iron skillet inside. Let the skillet sit in the oven for at least 20 minutes while you make the batter, so it gets hot.

Whisk together the oat flour, buckwheat flour, brown sugar, brown rice flour, salt, cardamom, baking powder, and baking soda in a medium bowl.

Blend the eggs in a blender on medium speed until pale and fluffy, about 1 minute. Add the milk and the dry ingredients and blend until smooth, about 30 seconds. You should have a well-combined thin batter, like a loose pancake batter.

When the batter is ready, open the oven door and add the butter to the hot skillet. Close the oven door and let the butter melt, about 30 seconds (it will foam and brown slightly). Carefully remove the skillet from the oven and quickly pour the batter into the center. Immediately return the pan to the oven and bake until the Dutch baby is puffed and crisp along the edges and golden brown in places, 12 to 14 minutes. Make sure you do not open the oven while cooking. By opening the oven door to check, you release the heat, which will prevent the Dutch baby from puffing up.

I like to serve this on a wooden board right on the table, with a kitchen towel or cloth napkin tied around the skillet handle, warning my guests that the pan is hot.

Tofu-Banana "Yogurt" with Market Fruit and Seed Cacao Crunch

Makes 4 servings

One of my family's favorite traditional desserts is bananas Foster. This riff on that midcentury New Orleans dessert is an easy breakfast dish to make ahead for friends or family, or even as a composed mini parfait dessert. It can be made with the yogurt of your choice, but even if you eat dairy, the silken tofu is a great alternative. Swap maple syrup or sugar for the date syrup if you like.

1 ripe or overripe banana

2 tablespoons date syrup, plus more as needed

Kosher salt

14 ounces (400 g) soft or silken tofu, drained

1 cup diced fruit of choice

Seed Cacao Crunch (recipe follows), for garnish

Warm spices, such as ground cardamom, cinnamon, or allspice, for topping (optional)

Preheat the oven to 425°F (220°C). Line a rimmed baking sheet with parchment paper.

Peel the banana and cut it in half lengthwise. Place the sliced banana flat side up on the prepared baking sheet, then drizzle with the date syrup and ¼ teaspoon salt. Bake for 15 minutes, untouched, until golden brown. The banana will break down slightly but shouldn't be melting. Remove the pan from the oven and let it cool slightly.

Transfer the caramelized banana, and any juices collected on the baking sheet, to a food processor or blender, add the tofu, and blend until fully combined, 15 to 20 seconds, scraping down the sides of the container halfway through. Taste and season with more date syrup or salt as desired. Move the mixture to the refrigerator to cool and stiffen up before eating, at least 1 hour. (This can be stored in an airtight container in the refrigerator for up to 1 week but is best eaten the day of.)

To serve, spoon some of the "yogurt" into a bowl and top with your favorite fruit, a couple tablespoons of seed cacao crunch, and a dusting of warm spices, if you like.

Seed Cacao Crunch

Makes about 1 quart (1.4 kg)

I use this over plain yogurt, as sprinkles for cookies, over ice cream, and more.

21 ounces (600 g) banana chips
10½ ounces (300 g) toasted
    pumpkin seeds
7 ounces (200 g) toasted pistachios
7 ounces (200 g) raw cacao nibs
1 ounce (30 g) black sesame seeds
1 ounce (30 g) white sesame seeds

Coarsely pulse each ingredient separately in a food processor, then sift each ingredient in a fine-mesh strainer to get rid of any dust. Combine all the ingredients and store in a jar at room temperature for up to 6 months.

There's an inspiring movement underway in this country centered on regional grains. An amazing beacon of that is Wild Hive Community Grain Project, whose grains we used at our restaurant for our sweet and savory porridge grains. Wild Hive and others like it are dedicated to creating high-quality grain products from the region where they are grown. They process stone-ground organic and heritage grains in small batches to provide the freshest whole-grain products, with a view toward creating a sustainable regional grain-based food system. I particularly love Wild Hive Farm's multigrain hot cereal mix, which combines winter wheat, spring wheat, rye, spelt, oats, corn, millet, and flaxseed; it's rich in fiber and protein and grown in a way that supports farm biodiversity.

Baked Porridge

Makes 4 servings

Porridge can get a soft reputation, but I love how it goes from morning to noon to night so seamlessly and, if prepared mindfully, can be packed with as much flavor as nutrition. It's not your mom's old-school oatmeal, that's for sure.

I make this often on weekend mornings with my family, as it's something easy that the kids can participate in preparing. I like to make a porridge bar, using a hodgepodge of little vessels and side bowls and surrounding my kids' plates with their choice of toppings so they can choose their own adventure. I've used regular milk here, but you can easily substitute hemp, oat, or your favorite nondairy milk.

5 ounces (150 g) multigrain hot cereal
   mix, such as Wild Hive Farm (see box)
1 vanilla bean, scraped
1 teaspoon sea salt
1 teaspoon ground cinnamon
½ teaspoon ground nutmeg
½ teaspoon ground cardamom
3 cups (720 ml) milk,
   plus more for serving (optional)
1 apple, pear, or stone fruit,
   cored and diced
2 tablespoons light coconut sugar,
   date syrup, or maple syrup
Optional toppings: fresh fruit slices,
   chia seeds, hemp seeds, nuts

Preheat the oven to 350°F (180°C).

Combine the cereal mix, vanilla seeds, salt, cinnamon, nutmeg, and cardamom in a large bowl. Stir in the milk and diced fruit, then pour the mixture into an 8-inch (20 cm) baking dish or cast-iron skillet. Bake for 30 minutes, until the oats are soft and the liquid is fully absorbed. Remove the dish from the oven and set the oven to broil.

While the broiler is heating up, sprinkle the coconut sugar or drizzle the syrup over the top of the cooked porridge. Broil for 2 to 4 minutes, until a crust has formed and the top is golden brown. Remove from the oven and serve warm in the baking dish, with any toppings you like and a little extra milk on the side, if desired.

Spring Pea Gazpacho

Makes 2 to 4 servings

As spring ripens into summer, this traditional pea soup–meets–gazpacho highlights the freshness of the season. Vibrant, green, crisp, and bright (aka everything we love about this time of year), this dish elevates any hot summer lunch or acts as a starter to a beautiful dinner. Typically, gazpacho features bread as a thickener (after all, *gazpacho* means "soaked bread" in Arabic). However, our twist relies on just peas and olive oil for its creaminess, which means it's naturally gluten free. A pistachio dukkah finish adds a salty crunch and a dose of healthy fats.

2½ cups (400 g) shucked fresh peas

2 tablespoons avocado oil

½ yellow onion, chopped

1 garlic clove, chopped

¾ cup (180 ml) water or
  Vegetable Stock (page 35)

2½ teaspoons sea salt, plus more
  as needed

¼ cup (25 g) micro pea tendrils

2 radishes, thinly sliced (I love using
  green or black radishes for this)

Juice of ½ lemon

2 to 3 tablespoons Herb Oil
  (page 40)

¼ cup Pistachio Dukkah (page 48)

Bring a medium pot of water to a boil over high heat and prepare a bowl of ice water on the side. Blanch the peas in the boiling water for 90 seconds, then transfer them to the ice bath to stop cooking and preserve their color. Remove the peas from the ice bath and place them in the freezer.

Heat 1 tablespoon of the avocado oil over medium heat. Add the onion and garlic and cook until soft, 1 to 2 minutes. Add the water, cover, and simmer until the flavors have combined, about 5 minutes. Remove from the heat.

Transfer the soup to a blender and puree until smooth (or use an immersion blender directly in the pot). Add the peas from the freezer and the salt and blend, gradually increasing the speed from low to high until the soup is creamy and smooth. If you want the soup to be thinner, add more water, a little at a time, and blend, repeating until you reach your desired consistency.

In a small bowl, toss the micro pea tendrils and radish slices with the remaining 1 tablespoon avocado oil, the lemon juice, and a pinch of salt.

Pour the soup from the blender into individual bowls. Garnish with the pea tendril salad, a drizzle of herb oil around the salad, and a sprinkle of pistachio dukkah for an added crunch.

Trumpet Mushroom Oyakodon

Makes 2 servings

An ideal one-pot dish with minimal cleanup, this oyakodon is often my go-to for a quick weeknight supper for two. If you can make eggs or an omelet, then you can easily re-create this. I love cooking in my donabe (we have quite a collection thanks to TOIRO) because it has such a subtle evenness to how it cooks and it also looks so beautiful when placed directly on a table to serve.

Oyakodon is a common Japanese comfort food, typically made with chicken. Here I use trumpet mushrooms, which have a delicate nutty flavor and a firm texture and are known to boost immunity and reduce inflammation (any mushrooms will do, though). It's an umami explosion and is warming, requiring not much on your grocery run. And it will get you making dashi to store in your freezer as a savory base for this and so many other dishes.

2 cups (480 ml) Vegan Dashi (page 37) or Vegetable Stock (page 35)

2 tablespoons dry sake or mirin

1 tablespoon soy sauce

1 tablespoon light coconut sugar or honey

1 small yellow onion, thinly sliced (about 4 ounces / 114 g)

12 ounces (340 g) mushrooms, thinly sliced lengthwise

3 scallions or ½ to 1 leek, trimmed and chopped

2 cilantro or shiso sprigs, leaves roughly chopped

3 large eggs, beaten

2 cups (400 g) cooked white rice (kept warm until ready to eat)

Toasted sesame seeds, Furikake (page 52), and/or togarashi seasoning, for garnish (optional)

Combine the dashi, sake, soy sauce, and coconut sugar in a medium skillet or donabe and bring to a simmer over medium heat. Stir in the onion and cook for 3 to 5 minutes, until the onion is slightly tender. Add the mushrooms and cook until all the vegetables are tender and cooked through and the broth has reduced by about half, about 5 minutes. Stir in half of the scallions and all of the herbs. Taste the broth and, if needed, adjust the flavor, sweet or savory, depending on your preferred taste.

Slowly pour the beaten eggs into the skillet and stir to incorporate. Cover the pan and cook until the eggs are just cooked but still loose, about 2 minutes.

To serve, ladle the egg mixture over bowls of warm rice and top with the remaining scallions. Toasted sesame seeds and furikake are wonderful additional toppings to garnish. If using a donabe, I serve from the pan with a side of rice and the garnishes on the table in little condiment bowls for each guest to serve themselves.

Kabocha Squash Curry Soup

Makes 4 servings

Kabocha is a winter squash originally farmed in Cambodia and now grown commonly in Japan. We grow it every season in our home garden and it just flourishes. Kabocha is a rich source of vitamin A, fiber, and beta-carotene, and it has a smooth, earthy flavor.

If you use another squash, you may or may not want to keep the skin in the soup, depending on if you care about the color. I love red kabocha squash because it's tender and you can keep the skin on, plus it has a savory, nutty sweetness, almost a cross between that of a sweet potato and a pumpkin. To me, this recipe blurs the line between sweet and savory. It is totally vegan, with natural creaminess from the squash and rich coconut milk (I love the Aroy-D brand).

1 red kabocha squash, 2 to 3 pounds (910 g to 1.4 kg)

1 head garlic, halved horizontally

4 tablespoons (60 ml) avocado oil

Sea salt and freshly ground black pepper

2 tablespoons curry powder

2 tablespoons ground cumin

1 tablespoon ground turmeric

1 small yellow onion, roughly chopped

2 (14-ounce/400-ml) cans coconut milk

Preheat the oven to 425°F (220°C).

Remove the stem from the squash, then cut in half from top to bottom. Scoop out the seeds with a spoon (they can be cleaned, dried, and fried up as a snack while you wait). Place each half cut side down on a rimmed baking sheet, along with the halved garlic head. Coat the squash and garlic with 2 tablespoons of the avocado oil and season with salt and pepper. Bake for 30 to 40 minutes, until both the squash and garlic are tender (a knife goes in easily and comes out relatively clean).

While the squash and garlic are in the oven, heat the remaining 2 tablespoons oil in a soup pot over medium-low heat until shimmering. Add the curry powder, cumin, and turmeric and cook, stirring, until slightly toasted and fragrant, about 30 seconds. Add the onion and cook slowly, mixing with a spatula every so often, until translucent. Remove from the heat and set aside until the squash and garlic are ready.

Remove the squash and garlic from the oven and place the squash in the soup pot over medium heat. Squeeze the garlic cloves out of their skin into the pot. Add the coconut milk and bring to a simmer. Blend the soup with an immersion blender, then season with salt and pepper. Remove from the heat and serve straight from the pot.

Store leftovers in an airtight container in the refrigerator for up to 7 days or in the freezer for up to 1 month.

Scrapped Cantaloupe Cucumber Soup

Makes 2 servings

I created this soup on the fly, using the scraps from the Melon and Cucumber Aguachile (page 120). It felt like a natural progression and an elegant way to upcycle, since I find that most recipes using cantaloupe don't utilize it all. Raw, refreshing, and a balance of sweet and savory, it mirrors the process and concept behind gazpacho. Remember, the idea here is to be flexible with the combination, since you're using leftovers. I love this soup as a bright starter on a warm day, or served in a small glass as an amuse-bouche for a cocktail party or to kick off a meal with a surprise-and-delight moment.

½ cantaloupe, rind and seeds
  removed, cut into large chunks
2 Persian cucumbers, trimmed,
  cut into large chunks
2 tablespoons avocado oil
Sea salt and freshly ground
  black pepper
Maldon salt
Marigold petals or microgreens,
  for garnish

Combine the cantaloupe and cucumber in a medium bowl. Add 1 tablespoon of the oil and season with sea salt and pepper. Blend using an immersion blender or a high-speed blender until smooth. No need to strain the soup, as the pulp helps the thickness and texture. Serve in the vessel of your choice, drizzled with the remaining 1 tablespoon oil. Add a pinch of Maldon salt and the marigolds or microgreens to finish.

To Continue

Melon as Itself

Makes 4 to 6 servings

My grandmother, who lived with me throughout my childhood, had melon every single day. It's a humble and often overlooked food, and I think that's why I love to cook with it so much. Plus it always reminds me of her and her simple morning routine of respite. In Los Angeles, the best melon you can find is from Weiser Family Farms, which is one of the most intentional, highest-quality farms in the country. The peak of melon season is summer, but you can try winter melon, pears, or even persimmons with this same preparation in the colder months. Choose a small melon for this recipe, as they tend to be sweeter and juicier.

1 cantaloupe or other melon

2 tablespoons fortified wine or vinegar

¼ cup (60 ml) avocado oil

2 ounces (55 g) aged Manchego cheese, thinly shaved into large pieces (about ½ cup)

Sea salt and freshly ground black pepper

Cut the melon in half horizontally and scoop out the seeds, saving any juice. Slice each half into six crescent moons about 1 inch (2.5 cm) thick (or thick enough to stand up on their own on the plate) and then carefully remove the skins using a sharp knife. Arrange the wedges on a serving plate, then cover them with the wine, oil, and cheese, in that order. Season with salt and pepper to taste.

PRO TIPS: Muscatel wine and chardonnay vinegar work best here. You can use any hard, sharp cheese in place of the Manchego, such as raw cheddar or Fiore Sardo.

Einkorn Salad with Charred Squash

Makes 4 servings

I made this salad on a whim when Camille Styles visited from Austin to host an intimate lunch in my garden, and it's one that I keep coming back to as a beacon of fall and the harvest moon. Einkorn is a delicious grain that's lower in gluten than regular wheat while being higher in protein and vitamins. Einkorn's protective hull allows the berries to be harvested without processing, and its deep root system doesn't require irrigation and thrives well in difficult seasonal and soil conditions, providing tremendous resilience compared to other wheat crops. The berries have a richness and a nuttiness, providing a hearty and sustainable base for any grain salad. Use whatever herbs you've got, such as parsley, chives, dill, mint, cilantro, or basil.

1 medium kabocha squash
Salt and freshly ground black pepper
2 cups (360 g) einkorn wheat berries
¼ cup (60 ml) avocado oil
Grated zest of 1 lemon
¼ cup (60 ml) fresh lemon juice
2 tablespoons honey
8 radishes, sliced
1 cup (20 g) chopped fresh herb leaves
¼ cup (50 g) pickled mustard seeds (see page 28)
¼ cup (30 g) toasted pine nuts

PRO TIP: I like Breadtopia and Revival einkorn from California, both of which can be purchased online.

Preheat a grill to high or the oven to 475°F (246°C).

Cook the kabocha whole by nestling it into the hot coals of a grill or roasting in the oven for about 25 minutes, until it's soft and tender throughout. When it's cool enough to handle, cut it down the middle and remove the seeds. Set aside.

Meanwhile, bring a medium saucepan of salted water to a boil. Turn the heat down to a high simmer, add the einkorn, and cook until al dente (tender but not mushy), 10 to 15 minutes. Drain, transfer to a serving bowl, and let cool.

To prepare the dressing, whisk the oil, lemon zest and juice, and honey in a small bowl until well combined. Season with salt and pepper to taste.

Add the radishes, herbs, mustard seeds, and pine nuts to the bowl with the einkorn. Season with salt and pepper, then add the dressing and toss to coat.

Use your hands to tear the kabocha into bite-size pieces, removing and composting the skin. Add to the bowl with the einkorn salad and gently toss. Taste to adjust the seasoning and serve.

Melon and Cucumber Aguachile

Makes 2 servings

This recipe was inspired by a massive cocktail event I cooked for, where I had to get creative with a very small kitchen. I had been asked to offer two or three dishes, but there was little space, and we could do almost no cooking because the hotel that hosted the event still had regular service running. As I have always said, perceived limitations breed the most brilliant inventions and creativity, and this dish is no exception.

I love the vibrancy of ceviche, but it usually includes seafood and you have to have the luxury of time. Ceviche requires a long marinade with acid from the lime juice to "cook" the raw seafood and tenderize the vegetables. By contrast, an aguachile is a quick bath in chili water to impart the pepper's flavor and it can be served immediately. The soft, delicate orange cantaloupe—particularly when it's juicy-ripe and in season—mirrors the delicate sweetness and texture of shrimp.

½ cantaloupe, chilled

2 Persian cucumbers

1 handful fresh basil

½ cup (120 ml) fresh lime juice

1 tablespoon honey or 2 tablespoons
  orange juice

1 jalapeño, seeds and ribs removed,
  finely chopped

Sea salt and freshly ground
  black pepper

2 tablespoons avocado oil

Trim the rind off the cantaloupe, leaving as much of the flesh as possible. Scoop the seeds right into a strainer set above a bowl so you can capture any juice from the center. Slice the cantaloupe in half crosswise, then cut thin wedges to resemble half-moon shapes (about 8 slices per piece). Set aside in the refrigerator to keep cold.

Using a mandoline, shave the cucumbers lengthwise to a similar thickness as the melon slices. (For the melon and cucumbers, you really can cut them any way you're comfortable, but there's something about this layered, thinly sliced look and texture that is a delicate complement to the punch of the spicy lime dressing.) Set the cucumbers aside.

Pick the basil leaves from the stems, keeping them together in small sprigs and saving the stems. Set aside.

In the bowl of a blender or food processor, combine the lime juice, honey, jalapeño, and basil stems. Season with salt and pepper. Process until smoothly combined. Transfer to a bowl, fold in the

cantaloupe and cucumbers, and gently stir until well dressed. Then, transfer everything to a plate, gently layering the melon and cucumber slices. Top with any extra melon juice left in the bowl and finish with a drizzle of oil and the basil leaves.

PRO TIP: I've garnished with melon cucumbers, one of my favorite little finds at my local market, but you don't need them. You could also top with thinly sliced pickled red onion.

Stone Fruit Carpaccio with Shaved Mushrooms and Roasted Figs

Makes 2 to 4 servings

As you can see from my recipes, I love an interplay of sweet and savory with an integration of fruits and vegetables throughout a meal, from starters through dessert. There's a vibrance and unexpected brightness when you start a meal with fruit in a savory style. Carpaccio of course is not just for proteins, and this dish delicately balances the hearty sweetness of ripe stone fruit with the woody aroma of mushrooms. I'd even think about throwing some Crispy Bulgur Wheat (page 50) on top for a last-minute crunch.

6 fresh figs, halved lengthwise
4 tablespoons avocado oil
1 teaspoon sea salt
1 large peach, halved and pitted
3 porcini or trumpet mushrooms, cleaned
1 tablespoon finely chopped fresh chives
Maldon salt and freshly ground black pepper

Preheat the oven to 350°F (180°C).

Place the figs skin side down on a rimmed baking sheet. Drizzle 2 tablespoons of the avocado oil over them and sprinkle with the sea salt, then roast until softened and slightly golden (they should not be falling apart, keeping their shape intact), about 15 minutes. Remove from the oven and set aside to cool slightly.

While the figs are roasting, heat 1 tablespoon of the remaining avocado oil in a cast-iron skillet over medium-low heat. When the oil is hot, add the mushrooms and cook for 2 to 3 minutes per side. You are not looking for color on these; you just want them cooked through lightly. Remove from the heat and set aside.

Using a mandoline, shave the peach into thin rounds. Place the pieces flat, overlapping slightly, across the whole bottom of a serving plate. Thinly slice the mushrooms and lay them delicately over the peach. Lay the warm fig halves across the carpaccio and drizzle the remaining 1 tablespoon avocado oil over the top. Sprinkle with the chives and season with Maldon salt and pepper to finish.

PRO TIP: If peaches are not in season, use whatever stone fruit you can find, just as long as it's juicy and ripe.

Cucumber and Green Plum in Its Juice

Makes 2 to 4 servings

I love to play with fruits and vegetables that share similar properties. This recipe highlights the interplay of cucumber and green plums, which share soft, porous texture, and a little bite from their skins, and meld together in a gorgeous verdant hue. The star on top is the intensely hot Thai chilis that prickle the gentle sweetness of the salad. Best served as cold as you can.

6 tablespoons (90 ml) fresh lime juice

¼ cup (60 ml) fresh tangerine or orange juice

1 tablespoon soy sauce

1 tablespoon peeled and grated fresh ginger

1 teaspoon honey

1 garlic clove, grated

2 Persian cucumbers, cut into ½-inch (1.25 cm) pieces

2 green plums, pitted and cut into ½-inch (1.25 cm) pieces

1 small shallot, thinly sliced

1 teaspoon coriander seeds

1 tablespoon Chili Oil (page 41) or 1 thinly sliced Thai chili (optional)

2 or 3 cilantro or Thai basil sprigs, leaves picked

In a medium bowl, whisk together the lime juice, tangerine juice, soy sauce, ginger, honey, and garlic until combined. Add the cucumbers, plums, and shallot and mix gently to coat the ingredients in the dressing. Set aside to marinate while you prepare the garnish.

Toast the coriander seeds in a small pan over medium heat, swirling the pan as they cook to promote even cooking. When the seeds become fragrant and slightly darkened, about 30 seconds, remove the pan from the heat and pour the seeds into a small bowl or plate to ensure they don't continue cooking and burn.

To serve, spoon the plum and cucumber mixture onto a rimmed plate or shallow bowl. Sprinkle the toasted coriander seeds over, then add the chili oil, if using. Finish with the cilantro.

Tomato and Stone Fruit Salad with Thyme Flowers

Makes 2 to 4 servings

There's something intriguing to me about foods that are the same, but not. Stone fruit and tomatoes share a color palette, texture, size, and aroma, while of course tasting completely distinct. This dish is peak summer, using different-colored tomatoes and fruit like green plums, apricots, or pluots to balance the sweet with the tart. For the cooler months, I'd even go for a similar mash-up with cucumbers and pears or even baked squash and apples. Let simple, pure ingredients dance with one another with little interference.

2 heirloom tomatoes

2 stone fruit, halved and pitted

Sea salt, to taste

2 tablespoons minced Preserved Lemons (page 30)

2 tablespoons extra-virgin avocado oil

1 flowering thyme sprig

Cut the tomatoes and stone fruit into like-sized wedges and combine them in a medium bowl. Season with salt and add the preserved lemon, then let them marinate together for 5 to 10 minutes. Arrange the tomatoes and stone fruit interspersed on a plate and finish with a drizzle of avocado oil and the flowers for garnish.

PRO TIP: Any edible flowers or herb flowers will work in place of the thyme flowers.

Beet Salad with Rye Crumble and Chili-Apricot Vinaigrette

Makes 4 servings

Salt baking is an age-old analog method that provides a blanket of insulation and allows for even, flavor-infusing, tenderizing cooking. The earliest recipe found for salt baking is from the fourth century BCE, in Archestratus's *Life of Luxury*. For me, salt baking is the answer for all root vegetables, and I love to combine salt-baked roots with some raw shavings of the same vegetable for a contrast of textures, temperatures, and techniques.

I'm not the biggest beet fan myself, but it's a top request when I cook for others. Our restaurant team made this dish for a beautiful private dinner for a fashion brand, and it left me wondering why I had resisted this jewel-like root vegetable for so long.

½ cup (60 g) walnuts, toasted and ground

½ cup (70 g) toasted rye breadcrumbs

7 medium beets (1 to 1¼ pounds/ 455 to 570 g)

1½ to 2 pounds (680 to 910 g) kosher salt (see Tip)

2 tablespoons avocado oil

Sea salt

1 cup (80 g) halved Sungold tomatoes

1 cup (120 g) strawberries, hulled and halved (if small you can keep them whole)

1 avocado, peeled, pitted, and diced (optional)

½ cup (120 ml) Chili-Apricot Vinaigrette (recipe follows)

Preheat the oven to 425°F (220°C).

In a small bowl, combine the walnuts and toasted breadcrumbs into a crumble, then set aside for garnishing later.

Scrub the beets and remove the greens (save for stock, a salad, pesto, etc.). Fill a roasting pan or cast-iron skillet with the kosher salt and bury 6 of the 7 beets, covering them completely. Roast for 1 hour to 1 hour 15 minutes, until a small knife inserted in a beet meets no resistance and comes out clean.

Remove the pan from the oven and, using a spoon, carefully crack the salt. Remove the beets from the salt and place them in a colander. Using gloves or a tea towel, carefully peel the beets while they are still hot (they will be easier to peel), gently rubbing the skin off and pulling it away from the flesh. (Alternatively, you can rinse the beets and let them cool to the touch, then peel them with a paring knife from the stem down.) Put the beets in a medium bowl and toss with the oil and a pinch of sea salt. Let rest for 5 to 10 minutes.

½ cup (6 g) fresh herbs, such as amaranth or opal or Thai basil, leaves picked, and/or edible flowers, such as nasturtium

While the beets are resting, slice the remaining raw beet thinly on a mandoline (no need to peel) and put directly into an ice bath.

Cut the cooked beets into ½-inch (1.25 cm) jewel shapes and put back in the oil to marinate for another 5 minutes.

In a separate bowl, combine the tomatoes, strawberries, and avocado (if you'd like). Drizzle with most of the vinaigrette and toss to combine.

Drain the raw shaved beets and dry well. Layer the raw beet rounds across the bottom of your serving plate, interweaving them starting from the middle to the outer edge. Arrange the chopped cooked beets and the tomato-strawberry mixture atop the shaved beets. Top with the walnut-breadcrumb crumble and drizzle the remaining vinaigrette around the plate. Finish with the remaining oil in the bowl and the herbs and flowers.

PRO TIPS: The salt from baking the beets can be reused as long as it's stored in an airtight container. I like red beets for this preparation, but you can use any kind you can find.

This goes so well with rye breadcrumbs, but truly panko or any leftover bread you want to make into crumbs will do; if you prefer it gluten free, extra toasted ground walnuts work too.

Chili-Apricot Vinaigrette
Makes about ½ cup (120 ml)

¼ cup (60 ml) extra-virgin avocado oil
3½ tablespoons (70 g) apricot preserves
1½ tablespoons Chili Oil (page 41)
1½ tablespoons fresh lemon juice
1 tablespoon minced shallot
¼ to ½ teaspoon kosher salt

Combine all the ingredients in a small bowl and whisk to combine.

The Whole Stalk or Bulb Salad

Makes 4 servings

I've always believed that learning to make great coffee sets the stage for learning to cook. A little bit of chemistry and a little bit of art, coffeemaking teaches the fundamentals of managing time, temperature, ratios, and heat.

Similarly, exploring how to compose a thoughtful, balanced salad is like getting on base. The architecture of a salad requires a base that can absorb the acid of the dressing. Usually this is lettuce, but as you see in this book, it can also be melon, celery, fennel, cucumbers, and the like. The idea is to get that base to absorb the acid and salt, almost like a quick pickle, and then coat the whole salad with oil at the end, trapping the flavor. In between, the key is to integrate sweet, sour, and bitter with an eye toward color and a variety of textures—having a crunch is essential. I also love to incorporate raw and cooked elements, with a mix of temperatures and textures. This allows you to bring the salad components together in your mouth and have an experience of discovery as you eat through it. It will reveal itself through the process and evolve as you savor it.

Fennel isn't for everyone. So, this same preparation also works well with whole celery, stalk to leaves. Another option is to use shaved celtuce, which is a sweet, tender Chinese lettuce that you can find at most Asian markets. For the smoked almonds, you can either purchase these online or from a specialty foods store, toast your own raw almonds with some smoked sea salt, or use regular toasted almonds.

2 ounces (55 g) blue cheese

¾ cup (180 ml) chardonnay vinegar

4 dates, pitted and chopped

12 ounces (340 g) fennel
(1 bulb with fronds) or celery
(4 or 5 stalks with leaves)

1 tablespoon minced Preserved
Lemons (page 30)

1 tablespoon brine from pickled chilis
(see page 28)

Sea salt and freshly cracked
black pepper

½ cup (70 g) smoked almonds,
toasted and chopped

¼ cup (60 ml) extra-virgin avocado oil

Put the blue cheese in the freezer for at least 1 hour and up to overnight, so it's hard enough to shave on a mandoline or with a cheese peeler.

In a small saucepan, bring the vinegar to a boil, then turn off the heat. Add the chopped dates, cover, and let them sit in the vinegar to rehydrate for 10 minutes (you still want them to have a bit of bite). Strain the dates over a small bowl, reserving the vinegar to be used in the dressing.

If using fennel, pick off the fronds and save for garnishing later. Using a mandoline or sharp knife, thinly shave the fennel, including the root end. If using celery, pick the leaves and save them for garnishing later. Using a mandoline or a sharp knife, thinly slice the entire stalk on a slight bias. Submerge the cut fennel or celery in an ice bath for 1 to 2 minutes to keep fresh and crunchy.

Drain the fennel or celery and dry with a kitchen towel or a salad spinner. Transfer the vegetable to a large bowl and add the preserved lemon, chili brine, reserved vinegar, and salt to taste. Let marinate for 3 to 5 minutes. Meanwhile, remove the blue cheese from the freezer and shave thinly onto a plate.

To serve, scatter the smoked almonds and dates on a serving plate. Arrange the fennel or celery in a small, delicate mound, top with the reserved fennel fronds or celery leaves and the blue cheese, then drizzle with the avocado oil and season with cracked pepper.

PRO TIP: I prefer Rancho Meladuco dates because they are the freshest and sweetest, and have a lovely fudgy texture.

Shaved Carrot Salad
with Mirin Vinaigrette

Makes 2 to 4 servings

One of my favorite New York restaurants in my old neighborhood is Cafe Gitane. I was always enamored of how they made a humble Moroccan-inspired shaved carrot salad such a star dish on the menu—one of the things I almost couldn't order a meal without. When I think about that dish, it comes together in my head as similar to Japanese sunomono salad, another staple that is relentlessly satiating. This recipe, done with two different humble ingredients, is an ode to and hybrid of both. It can be used as a slaw of sorts on tacos or over simple roasted vegetables or eaten on its own as a mainstay dish.

4 medium carrots, shaved thin with a
   peeler (about 2 cups / 245 g)
¼ cup (60 ml) Mirin Vinaigrette
   (recipe follows)
2 tablespoons thinly sliced fresh sorrel,
   basil, mint, or shiso
1 tablespoon avocado oil
1 teaspoon white sesame seeds or
   pistachios, toasted

In a medium bowl, toss the carrots with the vinaigrette and herbs. Transfer to a small serving bowl, drizzle with the oil, and top with the toasted seeds.

PRO TIP: Replace the carrots with 2 cups snap peas, sliced thin on a bias, to make a shaved snap pea salad.

Mirin Vinaigrette
Makes about ½ cup (120 ml)

This is one of my standard dressings that I keep in my refrigerator at almost all times. What I love most about it is how simple it is to make and that its versatility of flavor blends both sweet and savory. Drizzled over fruits and vegetables alike, it adds a rich umami and pop of vinegar to just about anything.

3 tablespoons toasted sesame oil
2 tablespoons rice vinegar
2 tablespoons soy sauce or tamari
2 tablespoons mirin
1 tablespoon honey

Combine all the ingredients in a bowl and whisk to emulsify, or put in a small jar and shake to emulsify. (One reason I like to use the latter method is so I can make a larger batch and store some in my refrigerator to use on the fly through the week).

Chef's Snack Celery Salad

Makes 2 servings

Celery is humble, ubiquitous, and inexpensive. It's a marshland plant harvested since antiquity that's almost entirely made of water. Its simple, gentle flavors are often overlooked as the star of a dish, yet most chefs I know adore this vegetable. It holds a tremendous amount of flavor when seasoned and composed properly, and in a certain order.

This dish is a tribute to the entire head of the celery and is intended to be created quickly, as chefs would do for a little pick-me-up snack while cooking on the line during service. Ideally, you are using a young head of celery here, which tends to be a little smaller, sweeter, and more tender. If you cannot find one, however, you can use half of a head of regular celery. If you really want to impress, see if you can find pink celery to make this recipe with.

3 tablespoons Chili Oil (page 41)
Juice of 1 lime
1 teaspoon Garlic Honey (page 46)
1 head young celery, leaves on,
   sliced on a thin bias
2 tablespoons Crispy Fried Garlic
   (page 45) or shallots
Maldon salt

Whisk together the chili oil, lime juice, and garlic honey in a medium bowl, then add the sliced celery and toss to coat. Let marinate for at least 5 minutes.

Spoon the dressed celery onto a plate or shallow bowl and pour over the remaining dressing from the bowl. Finish with the fried garlic and a pinch of salt.

Big Salad Energy

Makes 6 servings

An ode to European balance: there's nothing I love more than a deceptively simple salad as part of the meal. I often prefer to serve salad with the entree as a side dish, its acid-forward flavor a complement to whatever richer or more complex centerpiece stars on the table for the main meal. Meant to be a respite and easy delight, especially for those who may find side dishes to be too burdensome or daunting, this dish is also a great way to use up extra herbs and fronds.

A dear chef friend of mine gifted me a beautiful, prominent wood bowl that nestles on a three-pronged wood stand from Holland Bowl Mill, a multigenerational family-run business and one of the last (if not *the* last) wood mills in the country. The stand is the loveliest side piece and also provides a solution for never having enough space on the table. Plus, it inherently encourages guests to interact and serve one another, building intimacy and a shared experience from just a humble salad.

2 to 3 heads leaf lettuce (see Tip)
½ cup (120 ml) sherry vinegar
2 tablespoons Walnut Bagna Cauda (recipe follows)
Sea salt and freshly ground black pepper
1 cup (240 ml) extra-virgin avocado oil
½ cup (20 g) chopped fresh tarragon
½ cup (20 g) chopped fresh chives
½ cup (20 g) chopped fresh basil
½ cup (60 g) walnuts, toasted and chopped

Separate, wash, and dry the lettuce leaves. I like to keep them whole, but if the leaves are very large, you can tear or cut them into smaller pieces.

In a large serving bowl, combine the vinegar and walnut bagna cauda and mix with a fork until loosely emulsified, then season with salt and pepper. Layer in the lettuce leaves and gently toss until coated (there should not be a puddle on the bottom of the bowl, just enough to cover the leaves). Add the oil, toss to combine, and season with more salt and pepper if necessary. Sprinkle in the chopped herbs and walnuts and toss once more to combine.

Walnut Bagna Cauda

Makes 2 cups (480 ml)

Italians have perfected the ideal crudité dip with bagna cauda, which literally means "hot bath." The simple anchovy-based condiment is served warm with crisp, cold, raw and cooked vegetables. My vegetarian version uses nutrient-rich tahini and walnuts alongside preserves you are hopefully now storing in your kitchen—preserved lemon and white miso. If you can find red walnuts to try here, I highly recommend it.

½ cup + 1 tablespoon (125 g) tahini
¼ cup (95 g) Caramelized Onions
   (recipe follows)
6 tablespoons (90 ml) water
1 tablespoon (15 g) white miso
½ teaspoon toasted ground coriander
½ teaspoon ground saffron or turmeric
½ teaspoon ground Preserved Lemons
   (page 30)
½ cup (60 g) walnuts, toasted
   and chopped
¾ teaspoon grated lemon zest
2 teaspoons fresh lemon juice
Sea salt

Combine the tahini, caramelized onions, water, miso, coriander, saffron, and preserved lemon in a food processor and blend until smooth. Pour the contents into a small bowl and fold in the chopped walnuts and lemon zest, then season with the lemon juice and salt to taste. Leftovers can be stored in an airtight container in the refrigerator for up to 2 weeks. Bring back to room temperature to serve.

PRO TIP: I love Little Gem lettuce, escarole, red butter lettuce, or Castelfranco radicchio for the lettuce here, but if I could have Rosalba pink chicory all year round, it would be my dream. The crispness of the leaves and gentle bitterness pack so much flavor, and they just feel like gorgeous edible flowers. It's one of my favorite seasons of the year when they are popping. I love this compared to traditional lettuces particularly given there's such amazing variety across so many chicory species.

Caramelized Onions

Makes ¼ cup (95 g)

2 tablespoons avocado oil
2 large yellow onions, peeled and
    thinly sliced
Kosher salt
¼ cup (60 ml) broth, water, Vegan
    Dashi (page 37), wine, beer, or a
    sweet vinegar

PRO TIP: You can substitute shallots for
the onions.

Heat the avocado oil in a large pan over medium heat. Spread enough onions in the pan to create a flat single layer. Add a pinch of salt and toss to coat. Cook the onions, stirring constantly, until translucent, about 1 to 2 minutes. Add another single layer of onions with a pinch of salt, stirring and cooking for 1 to 2 minutes. Repeat until you've used all the onions.

Reduce the heat to medium-low and continue cooking, stirring occasionally and adding a bit of broth if the onions are browning too quickly or sticking to the pan. After 20 to 25 minutes, your onions should be a golden color. Deglaze the pan with the remaining broth.

Sungold Tomatoes and Marigold in Mirin Vinaigrette

Makes 2 to 4 servings

Sungolds are a chef obsession—I grow as much as I can in my home garden when they are in season. Not to be confused with cherry tomatoes (though they are an OK swap if you need to), they are juicy, bright, and oh-so-sweet with very delicate flesh, like pure sunshine in your mouth. Any edible flower or pretty herb with a neutral flavor can be swapped for the marigold.

2 cups (300 g) Sungold tomatoes, halved, chilled
Kosher salt
½ cup (120 ml) Mirin Vinaigrette (page 133)
¼ cup marigold petals (about 1 flower)

Arrange the tomatoes on a large plate from the middle out, cut sides facing up. Lightly salt the flesh and let marinate for 1 to 2 minutes. Drizzle the vinaigrette over the tomatoes and sprinkle with the marigold petals to finish.

PRO TIP: In the late summer or early fall months you can add husk cherries, which will blend right in and be a bit of a disguised surprise.

Shaved Kohlrabi with Cheddar, Pears, and Hazelnuts

Makes 4 servings

Kohlrabi is another incredible vegetable that's still lesser known. A cross between a turnip, a jicama, and broccoli stem, it's peppery and refreshing. You want to buy kohlrabi that is firm and without spongy spots and that has fresh-looking leaves. In this recipe, I prefer to use green kohlrabi to keep the salad looking monochrome, but purple will taste just as good.

1 green kohlrabi
¼ cup (60 ml) chardonnay vinegar
1 teaspoon kosher salt
1 ripe pear, cored and thinly sliced
¼ cup (20 g) thinly sliced Cabot white cheddar
¼ cup (30 g) hazelnuts, roughly chopped, toasted, skins removed
3 tablespoons avocado oil
Freshly cracked black pepper

Using a paring knife, trim away any flaws and remove the stems and leaves from the kohlrabi so you're left with a clean bulb (the leaves can also be used in anything you'd typically use kale or any other leafy green in, like salads or soups; any other scraps can be saved for your vegetable stock). Peel the bulb to remove the tough outer skin. Thinly shave the kohlrabi with a mandoline, cheese shaver, or knife and place in a medium bowl with the vinegar and salt.

Add the pear and gently mix to incorporate. Gently lay the salad mixture in a mound on a serving plate and drape the sliced cheese over the salad. Toss the hazelnuts in 1 tablespoon of the oil and sprinkle over the top. Drizzle the rest of the oil over the dish and finish with some pepper.

Charred Tokyo Turnips with Furikake and Whipped Mirin Tofu

Makes 2 to 4 servings

Tokyo (or Hakurei) turnips are a Japanese variety of this unassuming root vegetable prized for its high fiber, nutrients, and vitamins B6 and C. With over thirty types of turnips grown globally, this is one of the smallest, almost like a radish, and unlike most turnips, it can be eaten raw. That said, I love them charred with the greens on, arguably the best part.

I have a very fond memory of cooking this remotely on Zoom for an *Eater* series during the pandemic shutdown, with the notable natural winemaker Martha Stoumen. She was surprised by the ease and delicate flavor of this dish, which matched the earthiness of her thoughtfully low-intervention wines.

1 bunch Tokyo turnips, halved lengthwise, leaves left on

¼ cup (60 ml) avocado oil or other neutral oil, plus more as needed

Sea salt and freshly ground black pepper

1 (14-ounce/396-g) package soft or silken tofu, drained

3 tablespoons mirin

2 tablespoons ponzu sauce

1 tablespoon soy sauce

3 tablespoons Furikake (page 52 or store-bought)

Preheat the oven to 400°F (200°C).

In a large bowl, gently toss the turnips in the oil to coat evenly, then season with salt and pepper. Place each half turnip, cut side down, along the two short edges of a rimmed baking sheet, with the leaves facing into the middle of the pan, making sure they do not overlap. Roast for about 20 minutes, until the turnips are browned but not burned.

While the turnips are cooking, put the tofu in a blender or food processor and whip until smooth. Transfer the whipped tofu to a bowl and mix in the mirin, ponzu sauce, and soy sauce. If the mixture seems too thick to spread, you can thin it out a little by mixing in a drizzle of oil.

Transfer the roasted turnips to a bowl and toss with the furikake. Spoon the whipped tofu on a platter and place the furikake turnips on top. Serve right away, while hot.

Turmeric Tangerine Leeks

Makes 2 to 4 servings

Leeks happen to be one of my favorite vegetables—humble, inexpensive, and packed with balanced flavor. I am a deep fan of the allium, as a superfood centerpiece, with its inherent medicinal properties, including antimicrobial, antibacterial, and anti-inflammatory properties. Leeks are of the Earth, for the Earth, and for us.

This same preparation can be done using trumpet mushrooms, cut horizontally into rounds.

2 leeks
4 tablespoons (60 ml) avocado oil
1 teaspoon ground turmeric
¼ cup (60 ml) fresh tangerine or
    orange juice
Sea salt and freshly ground
    black pepper
3 pickled kumquats or calamansi
    (see page 28), thinly sliced
2 tablespoons basil oil (see page 40)
    or other herb oil
Maldon salt

Thoroughly clean the leeks. Trim off the dark green parts and the very bottom of the stem with the roots, all of which you can save for making stock. Slice the light green and white parts horizontally into ½-inch (1.25 cm) coins.

In a medium saucepan or cast-iron skillet, heat 2 tablespoons of the avocado oil with the turmeric until the spice blooms and becomes fragrant and toasted, about 2 minutes. Remove from the heat and carefully pour the oil into a bowl to cool to room temperature. Whisk in the tangerine juice and season with sea salt and pepper.

In the same pan, heat the remaining 2 tablespoons avocado oil over medium heat. Place the leek coins in a single layer in the pan and sauté until golden brown. Season with sea salt and pepper as they cook. Using tongs, carefully flip each leek coin to the other side, keeping its shape intact, and cook for another few minutes, until the other side is golden brown and the leek coins are tender. Transfer the leeks to a wire rack or towel to drain.

Arrange the warm leeks on a plate and spoon over the tangerine dressing. Drape the sliced pickled kumquats across the leeks. Drizzle the basil oil over the dish, and add a pinch of Maldon salt and pepper to finish.

Persimmon Caprese Salad

Makes 2 to 4 servings

In recent years persimmons have shifted from niche fruit to mainstream produce, and for good reason: they are packed with nutrients and vitamins, despite their small size. In the colder fall and winter months I think of and use persimmons as I would tomatoes. The squat-shaped Fuyu varietal turn bright orange when ripe and can be eaten firm or soft. The Hachiya varietal is more acorn-shaped and sweet only when they are ripe and soft; they are too astringent to eat otherwise.

You can use any bright, fresh herbs, like mint, basil, or pea shoots. I like to use white balsamic vinegar, but truly you can use any sweet vinegar or regular balsamic (reduced is even better). For this, I like keeping the colors of the ingredients pure.

3 ripe Fuyu persimmons
¼ cup (60 ml) white balsamic vinegar
Sea salt and freshly ground
    black pepper
1 (8-ounce/230-g) ball mozzarella
    or burrata
2 tablespoons extra-virgin avocado oil
¼ cup (3 g) herb sprigs
1 tablespoon Pistachio Dukkah
    (page 48, optional)

With a sharp knife, slice the persimmons thinly into rounds, being careful to cut around the top stem. Arrange the slices on a serving plate, overlapping, starting from the middle toward the outer rim. Drizzle with the vinegar and season with salt, then let marinate for 5 minutes.

Lay the cheese on top of the bed of persimmons, either whole to be dramatic or torn or sliced into pieces scattered around evenly. Season with salt and pepper. Drizzle the oil over the plate and top with the herb sprigs to garnish. For an optional crunch and twist, sprinkle over pistachio dukkah at the end.

PRO TIP: You can use a ripe Hachiya persimmon instead, but you will need only one since they're much larger. Make sure it's ripe to an almost jammy texture to ensure it's not starchy.

To Savor

Eggplant Katsu Sandwiches

Makes 4 servings

Katsu sandwiches are a casual staple in Japan, typically consisting of a breaded, deep-fried cutlet of chicken or pork served between slices of milk bread with a cabbage slaw and tonkatsu sauce (similar to barbecue sauce). Here, the hero ingredient is eggplant.

A nightshade fruit like tomatoes and peppers, eggplants have a low carbon footprint due to their efficient use of land and water through a long growing season, and they are grown broadly across the world. Eggplants are rich in antioxidants, vitamins B1 and B6, and a compound called nasunin that prevents cell damage in the body and reduces inflammation in the brain. Try graffiti eggplant if you can.

The eggplant here adds a tender, moist center and takes some inspiration from Italian eggplant parmesan with the mozzarella and confit tomatoes. The most important component is the traditional milk bread, shokupan, whose closest comparison is French brioche but which is lighter and made with far less butter. You can find milk bread at most Japanese grocery stores, or you can order loaves online from incredible Los Angeles baker Ginza Nishikawa.

1 large eggplant
½ cup (120 ml) soy sauce or tamari
½ cup (120 ml) toasted sesame oil
3 tablespoons honey
1 tablespoon nutritional yeast
1-inch (2.5 cm) knob ginger, peeled
1 or 2 garlic cloves
1 teaspoon chopped fresh rosemary
2 tablespoons kosher salt
1 cup (120 g) oat flour
3 large eggs
2 cups (120 g) panko breadcrumbs
½ cup (120 ml) avocado oil,
   plus more as needed
2½ ounces (70 g) mozzarella, sliced
½ head radicchio or cabbage

Preheat the oven to 400°F (200°C).

Trim the stem off the eggplant. Slice it lengthwise into ½-inch-thick (1.25 cm) pieces and lay them flat on a rimmed baking sheet. In a food processor or high-speed blender, blend the soy sauce, sesame oil, honey, nutritional yeast, ginger, garlic, rosemary, and kosher salt to make the marinade. Pour the marinade over the eggplant to cover and coat both sides of each eggplant piece, then let marinate on the baking sheet for 10 minutes.

Bake for 10 minutes, then remove from the oven and set aside to cool completely. Leave the oven on.

While the eggplant is baking, prepare your breading station: Put the flour in one bowl and

Juice of 1 lemon

Sea salt

8 slices milk bread, crusts on or off
depending on personal preference

2 ounces (55 g) preserved tomatoes

2 tablespoons tonkatsu sauce

2 tablespoons Dijon mustard
(optional, if you want a bite)

whisk the eggs in a separate bowl. Add the panko to a third bowl, and arrange the bowls in a line in that order. Set a wire rack on top of a rimmed baking sheet.

When the eggplant has cooled, bread it piece by piece by first coating gently with flour on each side, egg on each side, and then panko on each side. Set the breaded eggplant pieces on the prepared rack.

In a large skillet, heat the avocado oil over medium-high heat until hot but not smoking. Fry the eggplant in batches, being careful not to crowd the pan. Remove when golden brown on both sides, about 5 minutes, and place on the rack to drain. Replenish with avocado oil as needed and keep an eye on how quickly the pieces are browning, making sure to lower the flame if it seems they are starting to burn. Once all the eggplant has been fried, top each piece with a slice of mozzarella and place back in the oven just until the cheese melts. Remove the eggplant from the oven and let it cool slightly so you can handle it during sandwich assembly.

Shave the radicchio finely using a mandoline. Place in a mixing bowl and toss with the lemon juice and sea salt to taste. Let it sit to soften.

To build the sandwiches, take one slice of bread and spoon over some of the preserved tomatoes. Top with a couple pieces of the eggplant and cheese and then layer on some shredded radicchio. Drizzle tonkatsu sauce on another slice of bread, spread on the mustard, if using, and close the sandwich. Repeat with the remaining ingredients. Slice the sandwiches in half and serve.

Tartines

Makes 2 servings

Tartines represent the ultimate chef snack. On the line in the back of the kitchen, when you have just a fleeting moment to nourish yourself with whatever you have prepared or find lying around, the vessel is a crusty piece of bread to hold it all together to make it a meal, however improper. If you've ever worked in a restaurant, you know exactly what I mean.

Since half of this is bread, of course it's about the bread: fresh, splurge-worthy local artisan-bakery-type bread, thick cut and ready to be fried and slathered in the freshest ingredients you can get your hands on. Other than that, there aren't any rules. The key ratio is half gooey and half crunchy, and that applies to sweet or savory iterations. Slather as much or as little as you'd like on top, but to me, more is more. A few basic ideas are offered below to get you going, and then I encourage you to experiment—mix it up with the condiments, vegetables, techniques, and combinations we've explored together in this book. Tartines might look artful, but the goal here is straightforward: Get messy, no shame.

6 ounces (170 g) blue cheese

¼ cup (60 ml) avocado oil, plus more for drizzling

2 slices bread, cut thick (I recommend a simple country loaf or brioche.)

Maldon salt and freshly cracked black pepper

1 golden beet, roasted (see page 128) and sliced

¼ to ½ cup (60 to 120 ml) sheep's milk yogurt or crème fraîche

1 heirloom tomato, sliced

1 tablespoon toasted pepitas

Put the blue cheese in the freezer for at least 1 hour and up to overnight, so it's hard enough to shave on a mandoline or with a cheese peeler. When ready to cook, remove the cheese from the freezer and shave thinly onto a plate.

In a cast-iron skillet or other heavy pan, heat the avocado oil over medium heat. Fry the bread evenly on each side, moving it around and flipping it over as you go, so it's nice and crisped, and golden brown on each side. Remove from the pan and drain on a kitchen towel or a rack, hitting it with a pinch of salt while piping hot.

On one slice of bread, layer the beet slices, then top with the shaved blue cheese and a drizzle of avocado oil and season with salt and pepper. On the other slice, spread the yogurt, top with the tomato slices, and garnish with the pepitas. Season with salt and pepper and serve.

Maitake Mushroom Semolina Milanese with Shallot-Herb Labneh

Makes 2 servings

During our time in Rome, my husband and I fell in love with the city and the food as much as with each other. Milanese quickly became a favorite and core memory for our relationship and home going forward. This recipe adapts that technique for maitake mushrooms. Also known as hen of the woods, maitakes are a powerful adaptogen that reduces blood pressure and boosts immunity, and are often used in traditional medicine for their incredible healing properties.

The mycelium network of mushrooms is a key part of regenerative practices, as they help break down organic compounds that improve soil structure and release nutrients essential for plant uptake. Norwich Meadows Farm in upstate New York, a 250-acre farm cultivating over 1,500 types of produce run by Zaid and Haifa Zurdieh, is a cherished beacon for the New York City chef community. Smallhold is another supplier growing beautiful maitakes that I just can't get enough of. This recipe is perfect for a quick canapé or a family-style appetizer for any meal.

1 cup (240 ml) Shallot Labneh (page 91)

2 tablespoons finely chopped fresh flat-leaf parsley

1 to 2 heads maitake mushrooms

½ cup (80 g) semolina flour

3 large eggs

4 garlic cloves, grated

1 cup (140 g) breadcrumbs

6 tablespoons (25 g) grated pecorino cheese

1 to 2 cups (240 to 480 ml) avocado oil

Kosher salt and freshly cracked black pepper

1 lemon, cut into wedges

In a small bowl, mix the shallot labneh with the parsley and set aside.

Slice the mushrooms from top to bottom, keeping the core intact, to break up the whole maitake into pieces (each piece should be a wedge that can fit in the palm of your hand) so they hold together during frying.

Prepare your breading station: Put the semolina flour in one shallow bowl. In a second shallow bowl, whisk the eggs, then whisk in the garlic. In a third shallow bowl, mix the breadcrumbs and 2 tablespoons of the pecorino. Place an empty plate or baking sheet at the end of the line. Piece by piece, bread the mushrooms by first coating gently with the semolina flour on each side, then the egg on each side, and finally the breadcrumbs

on each side. Set the breaded mushrooms on the plate until ready to fry.

Heat the avocado oil in a large cast-iron or heavy-bottomed pan over medium heat until it is shimmering but not smoking. Carefully add the breaded mushroom pieces, one at a time, in batches if necessary, making sure not to crowd the pan. Fry until golden brown, then flip and fry the other side, 3 to 5 minutes. Once fully cooked, remove the mushrooms to a rack over a baking sheet or a towel to drain, and immediately season with salt.

When all the mushrooms are cooked, transfer them to a serving plate and top with the rest of the pecorino and some pepper. Serve with the shallot-herb labneh and lemon wedges on the side.

PRO TIP: If you don't have leftover bread at your house to make breadcrumbs, I also love panko breadcrumbs.

Pan-Roasted Leeks with Currants and Pistachio-Sorrel Gremolata

Makes 4 to 6 servings

To be honest, you'll want to eat these leeks even plain, but let's discuss gremolata. This condiment is typically a mix of oil, garlic, lemon, and parsley. I am obsessed with sorrel, which is such an underused, vibrant leafy green. In this gremolata, it brings a unique citrus note, but you can substitute dandelion greens, basil, tarragon, or parsley. Make the gremolata so it has at least 5 minutes to sit and meld flavors, but note that it is most vibrant and best eaten fresh, so don't let it sit for too long. Spoon over as much as you want on top of your leeks while they are still hot. You can plate it up, but I equally enjoy serving it right out of the skillet at the table.

Gremolata

1 tablespoon minced Preserved
    Lemons (page 30)
1 bunch (50 g) sorrel, chopped
1 tablespoon dried currants or any
    sweet dried fruit, like cherries
1 garlic clove, grated
1 tablespoon chopped salted pistachios
3 tablespoons avocado oil

Leeks

3 leeks
3 tablespoons avocado oil,
    plus more as needed
Sea salt and freshly ground
    black pepper
¼ cup (60 ml) chardonnay vinegar or
    sherry vinegar
1 tablespoon honey

To make the gremolata, mix the preserved lemon, sorrel, currants, and garlic in a small bowl. Fold in the pistachios and avocado oil and set aside.

Thoroughly clean the leeks. Trim off the dark green parts and the very bottom of the stem with the roots, all of which you can save for making stock. Slice the light green and white parts lengthwise into 6 batons.

Heat the avocado oil in a large cast-iron skillet over medium-high heat. Once hot, place the leeks cut side down in the pan and cook until golden brown, about 7 minutes, then flip each piece using a spatula or tongs. Season with salt and pepper. Add extra oil if needed, and when the other side is similarly golden brown, another 7 minutes, turn the heat down to medium-low and add the vinegar and honey to the pan. Be careful of flare-ups when adding the vinegar, as it may spray a little once it hits the hot pan, but lowering the heat before adding it will help. Continue cooking for about 10 minutes, until the vinegar and honey gently poaches and fully cooks the leeks.

Remove from the heat and transfer the leeks to a serving plate, cut side up. Spoon the gremolata over the leeks and serve.

PRO TIP: The gremolata is a great topping for any grilled or sautéed vegetables. I also love it spooned over frittatas or galettes.

Whole Roasted Romanesco
with Walnut Bagna Cauda

Makes 4 to 6 servings

While broccoli and cauliflower are mainstays, this is a recipe to get you to try something a little different that's essentially an artful hybrid of the two. Romanesco broccoli is as striking as it is delicious. It's bright when shaved and served raw, but here is a version of it cooked whole to be a showpiece on your plate. Just be mindful not to overcook it, as it should keep its shape and still have a little bite to it at the end.

1 whole head Romanesco, leaves left on
1 cup (240 ml) avocado oil
2 tablespoons ground spices,
    such as cumin, ras el hanout,
    and/or coriander
Sea salt and freshly ground
    black pepper
1 cup (240 ml) Walnut Bagna Cauda
    (page 136)
Chopped walnuts, breadcrumbs, or
    grated Parmigiano-Reggiano or
    pecorino cheese, for garnish
    (optional)

Preheat the oven to 425°F (220°C).

Place the Romanesco upright in a Dutch oven and pour over the avocado oil. Season with the spices and salt and pepper to taste. Cover and cook for 30 minutes, then remove the lid and finish roasting uncovered until golden brown, about 15 minutes more.

Transfer the Romanesco to a serving plate. Spoon the walnut bagna cauda on top and garnish with walnuts, breadcrumbs, or cheese, if desired.

Grilled Whole Escarole with Pine Nuts, Fiore Sardo, and Herb Oil

Makes 4 to 6 servings

This preparation seeks to strike a balance between a grilled salad and a roasted vegetable side dish. Escarole is a delicate chicory akin to an endive in flavor and a cross between kale and lettuce in texture. It's a versatile leafy green in the colder winter months that doesn't need much to sing on a plate, plus it offers fiber and vitamin A and supports bone growth, immunity, hormone balancing, and eye health.

Fiore Sardo is a delicate sheep's milk cheese originating in Sardinia, Italy. It's a bit gentler on the palate than pecorino. Feel free to use any hard, mildly briny cheese you like.

2 heads escarole, cleaned and dried
½ cup (120 ml) avocado oil
Sea salt and freshly cracked
　　black pepper
3 tablespoons chardonnay vinegar
½ cup (3.5 ounces / 100 g) thinly
　　shaved Fiore Sardo cheese
½ bunch fresh basil, leaves picked
¼ cup (30 g) pine nuts, toasted
¼ cup (60 ml) Herb Oil (page 40)

Cut each head of escarole lengthwise into four wedges, cutting straight through the core to ensure each wedge holds its shape and does not fall apart. Drizzle the avocado oil generously over each wedge and season bountifully with salt.

Heat a large cast-iron skillet over medium-high heat. Cook the escarole, in batches if necessary and turning only once, until the leaves are wilted and charred in spots, about 5 minutes on each side.

Transfer the escarole to a large bowl and, while still hot, season with the vinegar and pepper to taste. (The cold vinegar hitting the hot charred leaves will intensify the slightly smoky flavor.) Place the seasoned escarole on a serving plate and top with the cheese, basil, and pine nuts. Finish with the herb oil.

Wildfire Sweet Potatoes

Makes 4 servings

This dish is best done over raging-hot coals on an outdoor fire or grill, and not over high flames. For the potatoes, the bigger you can get, the better—think dinosaur eggs nestled in the embers. After cooking, the inside has a concentrated, almost candy-like flavor and a gentle, soft texture with smoky undertones from the burnt outer skin.

2 to 3 pounds (910 g to 1.4 kg)
  sweet potatoes, well scrubbed
½ cup (120 ml) extra-virgin
  avocado oil
Maldon salt

Prepare an outdoor fire or grill.

Place the sweet potatoes directly on the hot coals, leaving space between each sweet potato, and let them roast for roughly 1 hour, using long tongs to rotate them from time to time until all sides are completely black and charred. Check for doneness by poking the center of each sweet potato with a sharp knife: If it comes out clean and easily and the flesh feels softened, they are ready to go. Otherwise, continue cooking slightly longer.

Transfer the sweet potatoes to a serving platter. Carefully slice each one in half lengthwise, then drizzle generously with avocado oil and sprinkle with salt to finish.

Sherry Agrodolce Honeynut Squash with Fried Squash Seeds

Makes 4 to 6 servings

I love honeynut squash, a hybrid between winter and butternut squash, because of its smaller, individual size that minimizes waste no matter how small or large the group you're cooking for. I grow them in our garden from Dan Barber's Row 7 Seeds, and they are now quite ubiquitous in grocery stores during the fall and holiday season. I prepared this dish for a hundred people when I cooked at the Ecology Center in San Juan Capistrano, an incredible regenerative farm community anchor, so the recipe is designed to scale up for a large gathering. I often serve it in the baking dish, right out of the oven, for easy hosting. I like to keep the stem on for the whole look, and this recipe of course utilizes the seeds for garnish, with no element left behind.

Agrodolce is a Sicilian sweet-and-sour, sticky, molasses-like sauce. If you're short on time, store-bought pomegranate molasses is an excellent substitute. Montasio cheese is a beautiful cow's milk cheese from the Friuli region of Italy; you can substitute Gruyère or any semisoft, pungent option you can find (even blue cheese would be a fun kick on this).

4 honeynut squash
6 garlic cloves, peeled and smashed
1 bunch fresh sage, rosemary, tarragon, or fennel fronds
1 teaspoon ground cumin
1 teaspoon ground coriander
1 teaspoon freshly grated nutmeg
Kosher salt and freshly cracked black pepper
About 2 cups (480 ml) avocado oil

Preheat the oven to 375°F (190°C).

Cut the squash in half lengthwise (keep the stem on for presentation) and use a spoon to scoop out the seeds carefully so there's a nice spherical hole, saving the seeds for later. Place each half skin-side-down in a deep baking dish just big enough to fit all the squash. Scatter the smashed garlic over the squash, along with the herbs. Season with the cumin, coriander, and nutmeg, and salt and pepper to taste. Fill the baking dish with enough avocado oil so it goes up half of the height of the squash, and pour a little oil into each squash well on the flesh side (depending on

Fried Squash Seeds

Reserved seeds from 4 honeynut squash
¼ cup (60 ml) avocado oil
Kosher salt
1 teaspoon ground Aleppo pepper
   or Espelette pepper

Agrodolce

2 tablespoons avocado oil
1 shallot, minced
1 garlic clove, grated
1 cup (240 ml) sherry vinegar
½ cup (120 ml) honey
Sea salt and freshly ground black
   pepper

To Serve

½ cup shaved Montasio cheese
   (about 3.5 ounces / 100 g)

the size of your baking dish, you may need a little more or less oil). Cover with aluminum foil and bake for 25 minutes, until the squash is very tender but still holding its shape.

Meanwhile, clean and separate the seeds from the stringy flesh in a colander and then dry with a kitchen towel. Let the seeds dry on the kitchen counter for as long as you can. Heat the avocado oil in a cast-iron skillet over medium-high heat until it shimmers. Add the seeds, being careful not to crowd the pan, and cook, shaking and moving them around a bit, until they get golden brown. Remove the seeds from the pan with a slotted spoon and drain on a kitchen towel, hitting them with a pinch of kosher salt and the Aleppo pepper while piping hot. Set aside.

While the squash is cooking, make the agrodolce: Heat the avocado oil in a small saucepan over medium-low heat. Add the shallot and garlic and cook until soft but not brown, about 5 minutes. Add the vinegar and let cook down, reducing by three-quarters, about 5 minutes. Stir in the honey and let reduce further, until you have a thick, molasses-like consistency, 5 to 10 minutes. Season with sea salt and ground black pepper to taste. Some people add butter at this stage, but I leave it astringent, as is.

Assemble the cooked squash on a serving platter. Drizzle some of the cooking oil from the baking dish over the squash (and store the rest for anything, as it's infused with so much goodness), followed by the agrodolce. Sprinkle the cheese shavings over the squash and top with the fried squash seeds and cracked black pepper.

Fire-Roasted White Asparagus with Confit Shallots

Makes 4 servings

White asparagus is a gem of spring and summer, a perennial plant grown in a variety of climates without sunlight, which renders a milder, more tender composition than its green companion. I often get asked about how to cook it, as it can be a bit tougher and toothier if not done properly. This dish is about simplicity, grilling over fire, and lending that smoky char of the coals to the delicate asparagus. There's also just something about subtle monochrome dishes that's so enticing.

Sea salt and freshly cracked
  black pepper
1 bunch white asparagus
  (about 15 stalks)
3 tablespoons extra-virgin avocado oil
6 confit shallots (see page 45), halved
  lengthwise
2 tablespoons chardonnay vinegar
½ cup (3.5 ounces / 100 g) shaved
  Fiore Sardo cheese

Prepare a grill for direct heat.

Bring a pot of salty water to a boil and set up an ice bath in a large bowl. Once the water is boiling, blanch the asparagus for about 2 minutes. Remove the spears from the pot and immediately cool them down in the ice bath. Remove the asparagus and dry completely.

On a hot grill, char the asparagus spears lightly on all sides, being careful not to burn them. Transfer to a bowl and, while hot, toss with 2 tablespoons of the avocado oil, the confit shallots, and salt and pepper to taste. Transfer to a serving plate and drizzle the remaining 1 tablespoon avocado oil and the vinegar over the top. Top with the shaved cheese and finish with more pepper.

Charred Cabbage with Ricotta Salata, Hazelnuts, and Buttermilk Dressing

Makes 4 to 6 servings

My family's roots are in Eastern Europe, which I believe is where my passion for cabbage traces back to. Cabbage is a core ingredient in stews and soups as well as fermented condiments like sauerkraut and kimchi. Like root vegetables, cabbage is grown with low carbon emissions when compared to other foods. In season from fall to early spring, it's my accessible staple for the colder months. It's packed with nutrients like vitamins C and K, fiber, potassium, folate, and calcium. I love every kind of cabbage, and this recipe really can be made with any kind. It's particularly special when I can find purple napa cabbage or Caraflex cabbage at the farmers' market.

Buttermilk Dressing
¾ cup (180 ml) buttermilk
2 tablespoons crème fraîche
2 tablespoons avocado oil
1 garlic clove, grated
Grated zest and juice of 1 lemon
¼ cup (10 g) minced fresh chives
¼ cup (13 g) minced fresh tarragon
Sea salt and freshly ground
   black pepper

Cabbage
1 head green or purple cabbage
4 tablespoons (60 ml) avocado oil
Sea salt and freshly ground
   black pepper
4 tablespoons (56 g) unsalted butter
2 to 4 garlic cloves, peeled
¼ cup (60 ml) chardonnay vinegar

To make the buttermilk dressing, combine the buttermilk and crème fraîche in a large bowl. Whisk in the avocado oil, garlic, and lemon zest and juice, then fold in the chives and tarragon. Season with salt and pepper to taste, cover, and refrigerate until ready to use.

To prepare the cabbage, cut the head into eight wedges, cutting down through the core, ensuring that each wedge has a piece of the core attached. Coat all sides of the cabbage pieces using 2 tablespoons of the avocado oil, then season with salt and pepper.

Heat the remaining 2 tablespoons avocado oil in a large cast-iron skillet over medium-high heat. Place four of the cabbage wedges, cut side down, in the pan. Turn the heat down to medium-low, then add 2 tablespoons of the butter and 1 or 2 garlic cloves. Sear the cabbage, flipping once and basting with the butter every few minutes, until it's fork tender and charred, about 10 minutes. Transfer the cabbage to a bowl, then repeat this process with the remaining cabbage wedges, butter, and

To Serve

2 to 4 ounces (55 to 115 g) ricotta
   salata cheese, shaved
½ cup (40 g) chopped fennel fronds
½ cup (70 g) roughly chopped
   hazelnuts, toasted
Freshly cracked black pepper

garlic. Add the rest of the cabbage to the bowl, drizzle with the vinegar, and season with salt.

Place the cooked cabbage on a serving platter and pour the buttermilk dressing generously over the top. Garnish with the cheese, fennel fronds, hazelnuts, and some fresh cracked pepper.

PRO TIP: If you can't find ricotta salata, you can substitute another medium-hard cheese, such as white cheddar.

Al Pastor Squash Tacos

Makes 4 servings

Al pastor is my favorite taco, and in particular the ones from Leo's Tacos Truck in LA. There's much debate about who makes the best tacos in my hometown, but to me Leo's have been a landslide favorite for as long as I can remember. I don't have a trompo at home, but this is my homage to the nostalgic flavors that make Leo's tacos one of a kind.

The al pastor marinade is based on adobo, made from chilis, garlic, and vinegar, along with other aromatics, and of course the roasted pineapple is a must. The key is toasting the dried chilis before rehydrating them to make your paste. Adding cumin and dried oregano rounds out the flavors.

You've not lived until you've had Caramelo Sonoran-style tortillas from Lawrence, Kansas. We order them in bulk and freeze them until we are ready to use them. They take a few weeks to arrive since they are homemade, small batch, and made with love. I am especially fond of their avocado oil tortillas, every bit as luxurious and flaky as the ones made with animal fat. I also love to make my own tortillas with the incredible regeneratively grown heritage masa from Masienda, complete with their beautiful tortilla presses that you'll want to leave out on your counter for all to admire.

2 to 4 large dried guajillo chilis
1 tablespoon chili powder
½ cup (120 ml) Vegetable Stock
  (page 35)
3 tablespoons avocado oil
5 garlic cloves
2 tablespoons achiote powder
  or paste
1 teaspoon dried oregano
1 teaspoon ground cumin
¼ cup (60 ml) white wine vinegar or
  apple cider vinegar
¼ cup (60 ml) honey
2 teaspoons kosher salt

Preheat the oven to 400°F (200°C).

In a dry saucepan, toast the dried chilis and chili powder over medium heat until slightly darkened and fragrant, about 1 minute. Transfer to a bowl, cover with the stock, and set aside to hydrate for 5 minutes. When cool enough to touch, remove the hard stems from the dried chilis.

Heat 2 tablespoons of the avocado oil in the same saucepan over medium heat. Add the garlic achiote, oregano, and cumin and allow them to bloom for 1 to 2 minutes. Add the rehydrated chilis with the soaking liquid and cook for another minute. Add the vinegar, honey, and salt,

1 (1- to 2-pound / 455 to 910 g)
    winter squash, like honeynut,
    kabocha, or red kuri
1 pineapple
8 to 10 (5 inch / 13 cm) tortillas,
    warmed

Toppings

Finely chopped raw onion,
    pickled onions, or pickled shallots
    (see page 28)
Fresh cilantro leaves
Shaved radish
Lime wedges

then pour the mixture into a blender. Blend the marinade until smooth, about 2 minutes.

Cut open the squash and clean out the center flesh and seeds with a spoon, saving the seeds to toast and use at another time if you like (see page 161). Slice the squash into thin wedges and combine with the marinade in a large bowl. Let sit for about 5 minutes.

Trim the skin and top of the pineapple and cut in half lengthwise. Brush the outside of each pineapple piece with the remaining 1 tablespoon avocado oil and place cut-side-down on one side of a baking sheet. Arrange the marinated squash on the other side of the baking sheet, reserving the marinade, and bake for about 25 minutes. The squash will be a deep brown color and the pineapple should start to brown. Remove from the oven, and once cool enough to handle, thinly slice the pineapple and transfer to a bowl.

Spoon the squash onto the warm tortillas and arrange your toppings as desired: pineapple, onions, cilantro, radish, and lime wedges. Drizzle the tacos with reserved marinade and serve.

Braised Cabbage with Sour Beer Blanc and Toasted Pine Nuts

Makes 4 to 6 servings

This recipe is an ode to the Caraflex cabbage. Though still a somewhat experimental choice for farms, it's a hardy, economical crop that has a sweeter, milder flavor and texture than the traditional round cabbage. The tang of the sour beer alongside the floral sweetness of the orange blossom water brings a rich, vibrant roundness to the delicate cabbage leaves, topped with toasted pine nuts for a crunch.

1 (1- to 1½-pound / 455 to 680 g)
   head Caraflex cabbage
4 to 8 teaspoons kosher salt
4 tablespoons (55 g) butter
12 ounces (360 ml) sour beer
2 tablespoons honey
½ teaspoon orange blossom water
Freshly ground black pepper
½ cup (60 g) pine nuts, toasted
¼ cup (10 g) chopped fresh
   flat-leaf parsley

Preheat the oven to 350°F (180°C).

Quarter the cabbage, cutting down through the core to ensure that each piece has a piece of the core attached. Season the cabbage on all sides with 1 to 2 teaspoons salt for each wedge.

Heat a cast-iron skillet over medium-low heat and add 2 tablespoons of the butter. Swirl the butter around and, when it starts to foam, place the cabbage in the pan cut-side-down. Sear the cabbage until golden brown, then flip and sear the other sides, 3 to 5 minutes. Once browned, remove the pan from the heat, pour in the beer, and place the whole pan in the oven to braise for 10 to 20 minutes, or longer for larger pieces of cabbage. The cabbage should be softened through, which you can test by inserting a small knife through the middle of a wedge and meeting little to no resistance.

Transfer the cabbage to a serving platter. Place the pan with all the juices over high heat and add the remaining 2 tablespoons butter, the honey, and orange blossom water. Reduce this sauce, stirring constantly with a silicone spatula, until it starts to pull away from the bottom of the pan and is slightly thickened, 5 to 7 minutes. Season with salt to taste.

Season the cabbage with pepper, then spoon the sauce over the cabbage. Top with the pine nuts and parsley.

Roasted Vegetable Tagine

Makes 4 to 6 servings

Traditionally, tagine is baked in a tapered clay pot of the same name and captures the method and flavors of slow cooking something with healing spices (think saffron, ginger, cumin, cinnamon, paprika, chili), nuts, and dried fruits. The conical shape of the pot circulates steam as it cooks, with liquid dripping back down on the ingredients. It's a thoughtful, rich vessel for anything you have on hand and utilizes vegetables that you might otherwise have left over, unsure what to do with. And of course it can be served just as it's made, with minimal cleanup. The key here is using heartier vegetables that take a bit longer to cook, like beets, sweet potatoes, carrots, leeks, and winter squash. The dried fruit that works best is similarly more enduring—apricot, goldenberries, prunes, plums, cherries—to ensure all components keep their shape and bright flavor rather than breaking down too quickly during the long cooking process. My favorite dried fruits come from Frog Hollow Farm in Northern California.

4 tablespoons (60 ml) avocado oil
1 small yellow onion, diced
5 garlic cloves, grated
1 teaspoon peeled and grated fresh ginger
1 teaspoon ground turmeric
1 teaspoon ground cumin
1 teaspoon ground cinnamon
1 teaspoon coriander seeds
½ teaspoon chili flakes
Pinch saffron
2½ to 3 cups medium-diced hearty vegetables (see headnote)
1½ cups (240 g) cooked chickpeas
1 cup (240 ml) Vegetable Stock (page 35)
1 cup (88 g) halved grapes
½ cup medium-diced dried fruit

Heat 2 tablespoons of the avocado oil in a tagine or Dutch oven over medium heat until shimmering. Add the onion and cook for 5 minutes, until tender, then add the garlic, ginger, turmeric, cumin, cinnamon, coriander, chili flakes, and saffron. Bloom the spices in the oil for 1 to 2 minutes. Fold in the chopped vegetables and chickpeas and cook for about 5 minutes. Add the stock, grapes, and dried fruit and season with salt and pepper to taste. Turn the heat up to medium-high, cook for 10 minutes, then cover and turn the heat down to a gentle simmer for another 20 minutes, until everything is tender.

Finish by stirring in the preserved lemon and adjust the seasoning with salt and pepper if needed. Remove from the heat and sprinkle with the olives and remaining 2 tablespoons avocado oil. Serve with couscous, shallot labneh, and fresh herbs, if you like.

Sea salt and freshly ground
    black pepper
1 teaspoon minced Preserved Lemons
    (page 30)
½ cup (71 g) chopped pitted
    Castelvetrano olives

To Serve (optional)
Cooked couscous, rice, bulgur,
    quinoa, or bread
Shallot Labneh (page 91)
Fresh herbs, like mint, oregano, or
    fennel fronds

Roasted Trumpet and Maitake Mushrooms, Tatsoi, and Young Ginger Broken Rice

Makes 2 servings

I love finding ways to upcycle leftovers. This recipe transforms days-old rice into a luscious, pillowy, congee-like dish that's warming and vibrant. Young ginger has two times more polyphenols and antioxidants than mature ginger and has a more delicate floral flavor. The tatsoi here are like smaller flowering bok choy, adding an herbaceous crunch. This is a dish I've made for big events, a soothing lunch, and even a nourishing, savory breakfast.

½ cup (120 ml) Mirin Vinaigrette (page 133)

¼ teaspoon crushed Sichuan peppercorns (optional)

2 trumpet mushrooms, halved lengthwise

1 head maitake mushrooms, broken into florets

2 cups dashi or Vegan Dashi (page 37)

4 tatsoi (baby bok choy), quartered, with core intact

¼ cup (60 ml) toasted sesame oil

1 small yellow onion, finely chopped

1-inch (2.5 cm) knob young ginger, peeled and thinly sliced

2 garlic cloves, thinly sliced

1 stalk lemongrass, outer layer removed, thinly sliced into rounds

Sea salt and ground white pepper

2 cups (240 g) leftover cooked rice, any kind

Grated zest of 1 yuzu or lemon

Preheat the oven to 425°F (220°C).

In a medium bowl, combine the vinaigrette and Sichuan peppercorns, if using. With a knife, score the flat side of the trumpet mushroom halves. Add the trumpets and maitake florets to the bowl, toss with the vinaigrette, and let marinate for 10 minutes. Transfer the mushrooms to a small baking sheet and roast for 15 minutes. Remove from the oven and set on a plate to cool so they don't overcook.

While the mushrooms are in the oven, pour the dashi into a pot with a steamer basket or tray. Place the tatsoi in the steamer, cover, and steam over high heat until they are bright green and softened, about 10 minutes. Remove the steamer from the pot and set the tatsoi aside, but leave the dashi in the pot, covered, over low heat.

In a small skillet, heat the sesame oil over medium heat. Add the onion and sauté until translucent. Add the ginger, garlic, and lemongrass and cook until fragrant but not browned, about 2 minutes. Season with salt and white pepper to taste. Spoon in the rice and toss everything to coat. Slowly add the dashi from the steamer pot a little at a time, mixing it into the rice completely before adding more. You are looking for a final

To Serve

2 teaspoons avocado oil, toasted sesame oil, or Herb Oil (page 40)

2 teaspoons white sesame seeds, toasted

texture similar to risotto or congee. You may not need all the dashi, but if you've used all the dashi and need more liquid, you can add a little water as well. The rice is done when it feels cooked all the way but still with a slight bite, not mushy. It should also be spoonable and not too tight. Remove from the heat and fold in the yuzu zest.

Spoon the broken rice onto plates and arrange the mushrooms and tatsoi on top. Drizzle with the avocado oil and sprinkle with the sesame seeds.

California Pozole Verde

Makes 4 servings

Pozole (or posole in the southwestern US) is a controversial and heralded Mexican stew. With origins in Aztec sacrifice ceremonies, it features hominy, a preparation of nixtamalized dried corn. Heirloom and Indigenous strains of corn have been consciously preserved through seed saving, a centerpiece of food culture and the need to honor the past and important biodiversity through farming. Seeds are often considered critical for food sovereignty as well as local resilience to climate change.

I buy the Shabazi spice blend from La Boîte in New York, where chef Lior Lev Sercarz has been crafting his signature spice blends since 2006. Meant to mirror green zhoug, the blend includes cilantro, green chilis, coriander, and lemon juice. You can make your own or play with equal parts ground cumin, citrus dust (see page 47), and chili powder. For the tortilla chips, I usually make my own from leftover tortillas.

2 cups (450 g) dried hominy

3 tablespoons avocado oil

1 small yellow onion, quartered

2 garlic cloves, smashed

1 teaspoon Shabazi spice blend
  (see headnote)

Sea salt and freshly ground
  black pepper

10 cups (2.4 L) Vegetable Stock
  (page 35) or water, plus more
  as needed

¼ head (175 g) green cabbage,
  shredded

2 tablespoons chardonnay vinegar

1 tablespoon honey

10 tomatillos, husked and rinsed

1 poblano pepper

1 small jalapeño pepper
  (optional, if you like heat)

1 bunch scallions, trimmed

1 bunch cilantro

Grated zest and juice of 1 lime

Toppings (optional)

1 avocado, peeled, pitted, and diced

1 bunch red radishes with leaves,
  thinly sliced

1 cup (50 g) crushed tortilla chips

Lime wedges

Crème fraîche

The night before you want to make your pozole, soak the dried hominy in enough cold water to cover by 1 inch (2.5 cm) and let it sit on the counter overnight. Drain, reserving the soaking water.

Heat the oil in a large pot or Dutch oven over medium heat. Add the onion and garlic and sauté until tender, about 10 minutes. Season with the Shabazi and salt and black pepper to taste, and cook until fragrant, 1 to 2 minutes and the soaked hominy and cook until lightly toasted and golden. Add the hominy soaking water and the stock and simmer until the hominy is tender but not falling apart, about 45 minutes. Season with more salt and black pepper to taste.

Meanwhile, in a medium bowl, toss the shredded cabbage with the vinegar and honey and set aside to make a quick pickle.

Preheat a grill. Char the tomatillos, poblano, jalapeño, and 1 scallion together until they blister, brown, and are just beginning to collapse, 10 to 15 minutes. (Alternatively, you can char the vegetables on a rimmed backing sheet in a 425°F/220°C oven.) Transfer to a bowl to cool. When they are cool enough to handle, remove and compost the pepper stems and seeds.

Transfer the charred vegetables to a blender or food processor, add the remaining scallions, all but 3 or 4 cilantro sprigs, and the lime zest and juice. Blend until smooth and vibrant green. Add the puree to the pot with the hominy and bring to a simmer for at least 10 minutes; the longer you simmer, the more the flavors will meld and get richer. When you are ready, scoop out 2 cups (480 ml) of the soup and blend in a blender or food processor until smooth. Pour the puree back into the pot, mixing to combine (this will thicken the soup).

Serve the soup in bowls topped with the pickled cabbage, avocado, radishes (leaves as well), the remaining cilantro sprigs, and tortilla chips. Additional lime wedges and crème fraîche are great if you really want to do it up.

Heirloom Beans and Kale in Brodo

Makes 6 to 8 servings

Beans are an incredible superfood, and a tremendous vegetarian source of protein. My passion for beans started with my discovery of Rancho Gordo, a deeply mindful grower in Napa, California, dedicated to popularizing heritage beans that are native to the Americas. They ship nationwide and even offer a bean club to enthusiasts. I highly recommend exploring and experimenting with each of their varieties of beans, discovering new flavors along the way. Or you can start at your local farmers' market to try a type that's new. This is a humble preparation I make often for my family that is warming and soothing year-round.

1 pound (455 g) dried cranberry, yellow eye, pinto, navy, or cannellini beans
6 cups (1.4 L) water
5 tablespoons (80 ml) avocado oil
1 yellow onion, quartered
1 carrot, halved
1 celery stalk, halved
2 garlic cloves
1 bay leaf
¼ teaspoon coriander seeds
Sea salt and freshly ground black pepper
1 to 3 teaspoons chili flakes (optional)
1 cup (15 g) stemmed and shredded kale
Grated Parmigiano-Reggiano cheese, for garnish (optional)

About 12 hours before you want to eat, rinse the beans with cold water and check for small rocks or other debris. Put the beans in a large bowl, add the water, and let soak for 6 to 12 hours. (It's also totally fine to not soak your beans; it will just mean a slightly longer cooking time.)

Heat 2 tablespoons of the oil in a soup pot over medium heat. Add the onion, carrot, celery, garlic, bay leaf, and coriander seeds and sauté until the vegetables are tender and slightly golden, about 5 to 10 minutes, then season with salt, pepper, and chili flakes, if you want. Add the soaked beans and their soaking liquid (no need to waste the water, plus there's good vitamins and flavor in it). If needed, add more water to cover the beans by about 2 inches (5 cm). Bring to a rapid boil for 10 to 15 minutes, then turn the heat down to a gentle simmer.

The beans will take about 1 hour to cook if they were soaked, or 2 to 3 hours if not soaked. If the water gets low during cooking, replenish with boiling water as needed; the beans should remain submerged. Once the beans are tender, remove the pan from the heat, cover, and let

stand for 30 minutes to 1 hour. Season with salt and pepper to taste and stir in another 2 tablespoons oil.

There should be ample broth surrounding the beans. Bring it up to a simmer, add the shredded kale, and let cook lightly in the hot broth. Drizzle over the remaining 1 tablespoon oil, some more pepper, and cheese to finish, if you like.

PRO TIPS: Rule of thumb: 1 cup (250 g) dried beans yields about 3 cups (600 g) cooked beans, so 1 pound dried beans (which is about 2 cups) will yield about 6 cups (1,200 g) cooked beans.

In this recipe, you're cooking the beans and making a vegetable stock at the same time. If you would rather use a homemade Vegetable Stock (page 35) or a store-bought stock, you can skip the carrot, onion, celery and bay leaf and go straight to boiling the beans in the vegetable stock.

Rigatoni with
Next-Day Squash Bolognese
Makes 2 to 4 servings

For this recipe, I upcycle cooked squash from a prior meal, but you can certainly use fresh diced squash. You can use any hard squash, but I don't recommend unpeeled delicata squash; the skin tends to remain hard throughout cooking. If using skin-on squash, blend with an immersion blender at the end, just to break down those peels. If using peeled squash, it should break apart during cooking and won't require blending. The longer it cooks, the better.

¼ cup (60 ml) avocado oil

1 cup (120 g) finely chopped celery

1 cup (140 g) finely chopped yellow onion or shallot

1 cup (140 g) finely chopped carrot

2 garlic cloves, grated

2 tablespoons tomato paste

2 cups (250 g) small-diced squash (see headnote)

1 Parmesan rind (optional)

4 cups (945 ml) Vegetable Stock (page 35), plus more as needed

2 tablespoons kosher salt

8 ounces (230 g) rigatoni

½ cup shaved Parmigiano-Reggiano, ricotta, or mascarpone cheese (optional)

In a large saucepan, heat the avocado oil over medium heat until it shimmers. Add the celery, onion, carrot, and garlic and sauté until light in color but not browned, 5 to 7 minutes. Add the tomato paste and mix to break it apart. Add the squash and lightly cook for 3 to 5 minutes. For added viscosity and flavor, add a Parmesan rind if you have one on hand.

Add the stock, turn the heat down to medium-low, and simmer the sauce for about 1 hour. If you notice the sauce starting to dry out, you can add a little more stock or some water to keep it hydrated while you continue cooking it down. When you notice the sauce is thickening and close to being done, fill another large pot with water, add the salt, and bring to a boil over high heat. Add the pasta and cook until al dente according to the instructions on the package.

If you used skin-on squash, lightly blend the sauce with an immersion blender to help break everything down. If you peeled the squash, use the back of a spatula to break apart any larger pieces. If you are adding cheese, fold it into the sauce. Taste and adjust the seasoning one last time, then drain the cooked pasta and add it to the sauce, mixing lightly to combine.

Brothy Ditalini and Chickpeas

Makes 2 to 4 servings

This recipe is an ode to the classic holiday dish tortellini en brodo, from the Emilia-Romagna region of Italy. The famed preparation is particularly treasured in Modena, where I visited on my babymoon, pregnant with our first child. We had the privilege of staying in the newly opened Casa Maria Luigia, owned by Massimo Bottura and Lara Gilmore, and Chef Massimo's version of this heritage soup brought me back to a different era entirely. There's not much to it; the beauty is in the simplicity and purity of the ingredients, with gentle seasoning to balance it all out.

4 tablespoons (60 ml) avocado oil

1 small yellow onion, thinly sliced

4 garlic cloves, sliced

1-inch (2.5 cm) knob ginger, peeled and grated

1 (15.5-ounce/440-g) can chickpeas, undrained

1 cup (240 ml) Vegetable Stock (page 35)

Sea salt and freshly ground black pepper

½ cup (60 g) ditalini or conchigliette

Heat 2 tablespoons of the avocado oil in a medium pot over medium heat. Add the onion, garlic, and ginger and lightly sauté until translucent but not browned at all. Add the chickpeas and their liquid and the stock and bring to a simmer. Season with salt and pepper to taste, then add the pasta. Simmer until the pasta is al dente, according to the instructions on the package. Remove from the heat, ladle into bowls, and drizzle with the remaining 2 tablespoons avocado oil.

Whole Zucchini Bucatini

Makes 2 servings

A dish that could very well be my dream last meal is from Lo Scoglio on the Amalfi coast of Italy, a family-run restaurant casually situated on the water just off a small dock. While they are known mostly for their self-caught fresh seafood, their zucchini pasta is what I dream about. My version is centered on the whole zucchini, ideally using baby zucchini with the flowers on, if you can find them. Using every single element of this gorgeous vegetable gives additional color and floral aromas to the finish of the dish. I usually double the recipe to save some for later; it's as delicious cold, the day after. You can use any pasta shape you like here.

Sea salt and freshly ground
   black pepper
8 ounces (230 g) bucatini
½ cup (120 ml) avocado oil
12 baby zucchini, sliced into
   ¼-inch-thick (6 mm) coins
Juice of ½ lemon
Squash blossoms, stamens gently
   removed, for garnish
1 cup (30 g) grated
   Parmigiano-Reggiano cheese

Bring a large pot of salted water to a boil and cook the pasta until al dente according to the package instructions.

While the pasta is cooking, heat the avocado oil in a large, wide saucepan over medium-high heat. Add the zucchini, season with salt and pepper, and cook until slightly browned and cooked through but still holding their shape, about 1 minute on each side. Remove with a slotted spoon and drain on a rack or kitchen towel. Transfer the zucchini to a large bowl and add the lemon juice.

Once the pasta is cooked, scoop out about 1 cup (240 ml) of the pasta water and drain the pasta. Add the pasta to the bowl with the zucchini and add the pasta water in little increments, to incorporate the fried zucchini into a sauce (the starch in the water will help that along).

Divide the pasta between two plates and top with the squash blossoms, cheese, and a final sprinkle of salt and pepper.

Spaghetti with Tomato and Raw Corn

Makes 4 servings

This is summer in a bowl for me, particularly when I can get Sungold tomatoes, one of my favorite pieces of produce. You can make this with preserved tomatoes, but you will have to adjust the salt accordingly, as preserves are presalted. The miso gives a hint of umami, and the raw corn lends a fresh, crisp crunch and bright creaminess to the classic pasta pomodoro.

2 ears corn
4 tablespoons (60 ml) avocado oil
2½ pounds (1 kg) Sungold tomatoes
1 teaspoon white miso
Sea salt and freshly cracked
   black pepper
1 pound (455 g) spaghetti
2 basil sprigs, leaves picked
½ cup (15 g) grated
   Parmigiano-Reggiano cheese

Remove the husks and silks from the corn (save for stock), cut the kernels off the cobs, and set them aside in a bowl.

In a large saucepan, heat 3 tablespoons of the avocado oil over medium heat. Add the tomatoes and cook until they start to burst, about 10 minutes. Turn the heat down to a simmer, gently stir in the miso, and cook for 5 minutes, until the tomatoes start to break down into a sauce. Season with salt and pepper to taste.

While the sauce is cooking, bring a large pot of salted water to a boil. Gently lower the spaghetti into the pot, being careful not to break the noodles but ensuring they are submerged. Cook the pasta until al dente according to the instructions on the package. Scoop out ½ cup (120 ml) of the pasta water, then drain the pasta.

Immediately add the pasta to the tomato sauce and turn the heat up to medium-high. A little bit at a time, add the pasta water, which will help thicken the sauce and make it stick to the pasta. You know you have stirred in enough pasta water when the pasta sauce turns glossy. Toss in half of the corn kernels and remove the pan from the heat.

Twirl the pasta into bowls to serve. Top with the remaining corn, the basil leaves, cheese, remaining 1 tablespoon avocado oil, and more pepper to finish.

Lemon and Red Lentil Risotto with Cured Egg Yolk

Makes 2 servings

While dining with Cal, the proprietor at Thacher House in Ojai, I had a version of this dish that was so pure and beautifully simple. Although there were few ingredients, it felt like it warmed my whole body, top to bottom. I absolutely love lentils, and since they're a key cover crop that's also a high-protein legume, I hope to bring them to the fore as the centerpiece of a table. So, here is my spin on that memorable dish in the style of risotto, with an added kick of brine from either a cured egg yolk or Mimolette cheese. You could also incorporate nutritional yeast when you do the final seasoning in lieu of the egg yolk or cheese.

Kosher salt

1 large egg (optional)

3 tablespoons avocado oil

1 small yellow onion, finely chopped

1 cup (225 g) red lentils

Freshly cracked black pepper

3 cups (710 ml) unsalted Vegetable Stock (page 35), warmed

Grated zest and juice of 2 lemons

Sea salt

2 ounces (55 g) Mimolette cheese (optional)

If using cured egg yolk, start the curing process 2 days before. Fill a container with a layer of kosher salt. Separate the egg yolk from the white (save the white for another use) and place the yolk on top of the salt. Cover the yolk with more salt, place the lid on the container, and let it cure in the refrigerator for 2 days. Before using, remove the yolk from the salt and lightly brush off any excess salt.

Heat the oil in a medium saucepan over medium heat. Add the onion and sauté until translucent but not brown at all, about 3 minutes. Stir in the lentils to coat with the oil and onion and cook for about 5 minutes. Season with ¾ teaspoon kosher salt and a pinch of pepper.

Stirring constantly, add the warm stock ½ cup (120 ml) at a time, waiting until it's fully absorbed before adding more, 15 to 25 minutes. The lentils should be completely split open, creamy, and tender, about the consistency of risotto. Remove from the heat and stir in the lemon zest and juice. Add sea salt and more pepper to taste.

Divide the risotto between two bowls and finish by grating the cured egg yolk or cheese over the top, if using.

Barley with Broccoli Pesto

Makes 2 servings

Barley is as powerful for our soil as it is for our health. It is an anchor of crop rotation for regenerative farms, acting as a natural weed repellent that improves soil structure and helps reduce erosion. Plus, it's high in vitamins, minerals, and fiber, making it far more nutritious than other grains like wheat.

This is a quick and easy one-pot recipe inspired by my favorite broccoli pasta from Cul de Sac in Rome, from when I studied abroad in Italy twenty years ago. The key is blanching the whole broccoli (stem and all) in salted water and then using that same water to cook the barley (as you would pasta). Simple blending of the sauce with an immersion blender and sautéing in ample lemon juice and zest gives it a bright finish to complement the creamy cheese and heartiness of the barley. Simple, conscious, and nourishing vibes.

Sea salt and freshly ground
   black pepper
1 (1-pound/455-g) head broccoli,
   stem and florets cut into
   small chunks
1 cup (200 g) pearl barley
3 tablespoons avocado oil
Grated zest and juice of 1 lemon
¼ cup (8 g) grated
   Parmigiano-Reggiano cheese

Fill a medium pot three-quarters full with water and season with a few tablespoons of salt (the water should taste salty). Bring to a boil, then add the chopped broccoli and blanch until soft, 7 to 10 minutes. Using a slotted spoon, transfer the broccoli to a medium bowl.

Bring the water back to a boil, add the barley, and cook, stirring occasionally, until tender with a slight bite, 25 to 30 minutes. Ladle about 1 cup (240 ml) of the cooking water into the bowl with the broccoli and drain the barley, discarding the remaining cooking water.

Heat the avocado oil in the same pot over medium heat. Add the broccoli and reserved cooking water and cook down until falling apart, roughly 10 minutes. If you like, you can blend this into a smoother sauce at this point using an immersion blender. Add the cooked barley and stir to coat. Remove from the heat and stir in the lemon zest and juice and cheese to finish.

Eggplant Green Curry with Cardamom Rice

Makes 4 servings

Curry has been a recent culinary exploration for me. I didn't grow up with a traditional family preparation, and long found it daunting to make, as it seemed too nuanced to make me feel confident. It was during the pandemic shutdown that I threw myself into learning. Until Justin Pichetrungsi of Anajak Thai in Los Angeles—one of my favorite restaurants—teaches me the secret of his family's panang curry, this green curry has my heart. Though I would love to make my own curry paste (maybe that's for my next book), I've tested many store-bought versions, and I find the best results with the Maesri green curry made with fresh green chilis, lemongrass, galangal, garlic, shallots, and other herbs and spices. If you have any trouble finding it, you can order it online from the best specialty store in New York City, Kalustyan's.

This is a one-pot dish I make quite regularly. It's also a beautiful large-format dish for gatherings; I love to serve it out of a massive clay pot tableside. The sourcing of the right high-quality ingredients will make a difference to the flavor in the end. My favorite rice brand is Lotus Foods, which is a pioneer in regenerative practices for rice growing. Lundberg Family Farms is another brand that has Regenerative Organic Certification and can be found in most major grocery chains.

PRO TIPS: Grind the green curry paste and Thai basil leaves together before adding to the curry for enhanced color.

I recommend Aroy-D coconut milk because it has more richness and less water content than others. This brand is also the only one I've found that's BPA free and does not have stabilizers, thickeners, gums, or preservatives.

Rice

1 cup (200 g) white rice
1 teaspoon kosher salt
½ teaspoon ground cardamom
½ cup + 1 tablespoon (135 ml) water
½ cup + 1 tablespoon (135 ml)
  coconut milk

Curry

2 ounces (55 g) green curry paste
1 Thai basil sprig, leaves picked, stem
  reserved
1 large eggplant, trimmed
Sea salt and freshly ground
  black pepper
¼ cup (35 g) rice flour
6 tablespoons (90 ml) avocado oil
1 stalk lemongrass, outer layer
  removed, finely chopped
½ yellow onion, thinly sliced
1-inch (2.5 cm) knob ginger,
  peeled and grated
1 garlic clove, grated
2 (8.5-ounce/510 ml) cartons
  coconut milk
4 cups (1 L) Vegetable Stock
  (page 35)
1 tablespoon soy sauce or tamari
1 tablespoon honey
1 sprig curry leaves
1 or 2 Thai chilis (optional)

To Serve (optional)

Lime wedges
Coconut yogurt or labneh
Fresh mint or pea shoots

To make the rice, in a strainer set over a bowl in the sink, rinse the rice four times under cold water until the water runs clear. The fourth and last time, fill up the bowl and let the rice soak in the water for 20 minutes, then drain. The key ratio for cooking rice is 1.1 portion liquid to 1 portion rice. Put the soaked and strained rice into a rice cooker with the kosher salt, cardamom, water, and coconut milk and cook according to your rice cooker's instructions. Keep the rice warm while you make the curry.

Grind the curry paste and Thai basil leaves in a mortar and pestle and set aside.

Cube the eggplant, put it in a colander in the sink, sprinkle with sea salt, and drain for 5 minutes. Once the moisture has been released, coat the eggplant evenly with the rice flour, shaking off any excess. In a Dutch oven, heat 3 tablespoons of the avocado oil over medium heat. Add the eggplant and sauté until golden brown on all sides and cooked through, 7 to 10 minutes. Transfer the eggplant to a bowl and set aside.

Add the remaining 3 tablespoons avocado oil to the pot and bring back up to temperature. Add the lemongrass and sauté for about 5 minutes. Add the onion, ginger, and garlic, season with sea salt and pepper, and sauté until a little golden, about 5 minutes. Add the curry paste mixture, allowing it to infuse in the aromatics for about 2 minutes. Return the eggplant to the pot. Cover everything with the coconut milk and stock and bring to a boil, then reduce to a simmer. Once the broth is gently simmering, add the soy sauce, honey, Thai basil stem, and curry leaves. Simmer the curry until thickened, about 20 minutes, then season with sea salt and pepper if needed. Remove the Thai basil stem and curry leaves and, if desired, add the chilis.

Serve the curry over the rice, along with lime wedges, yogurt, and extra herbs if desired.

To Finish

Dark Chocolate Bark with
Bee Pollen, Rose Petals, and Pink Salt

Makes 6 servings

I grew up with Rocky Mountain Chocolate, as my dad is originally from Denver, Colorado. Especially during the winter, the chocolate was our treat and something we patiently looked forward to when we went to his home state and got to savor all of the handcrafted creations. My parents' particular favorite was the dark chocolate bark, often with almonds but sometimes different combinations of whatever was in season. At the time, of course, this was too grown-up for me, paling in comparison to my favorite: chocolate-dipped honeycomb.

This recipe recalls the bark of my childhood and combines a number of ingredients I love, like bee pollen, while being both a vibrational and indulgent dessert. Bee pollen is a superfood—it lowers cholesterol, reduces hardening of the arteries, and improves metabolism—but I hear often from friends that it feels like a mystery to use. Here it provides a crunch and complements the floral notes of the rose. You can also have some fun with these toppings, using anything that offers some crunch and flavor in a small, sprinkleable size. I'm a magnet for large-format anything, especially dessert, and this is a deceptively simple yet notable centerpiece for a table that comes with an irresistibly fun activity: instead of breaking bread, you can break bark. Most importantly, invest in a mallet so you can have fun smashing the bark.

20 ounces (570 g) dark chocolate (see Tip)

3 tablespoons bee pollen

2 tablespoons crushed dried rose petals

2 tablespoons pink Himalayan salt or Maldon salt

Line a small rimmed baking sheet with parchment paper and set aside, making sure you have enough room in your freezer to fit the tray later on. Set up a double boiler on your stovetop to melt the chocolate or improvise one by setting a metal mixing bowl over a saucepan containing about 3 inches (7.5 cm) water, and bring to a boil. Put the chocolate in the upper pan or bowl and swirl it with a silicone spatula just until fully melted. Remove from the heat, then carefully and

quickly pour the chocolate onto the prepared pan, gently spreading it evenly with the spatula. The goal is for the bark to be about ¼ inch (6 mm) thick. Once the chocolate is laid down and still soft, sprinkle the rest of the ingredients evenly and as desired. Place the tray in the freezer for at least 1 hour or ideally overnight. When ready to serve, bring the chocolate to about room temperature and break into pieces to serve.

PRO TIP: There has been a strong movement toward reforming chocolate production so that it better cares for the farming ecosystem with the soil nutrition in mind, as opposed to overcultivation of this monocrop to address rising demand. For the lowest-impact chocolate, choose Fair Trade, Fair for Life, non-GMO, and organic certifications; buy from brands who know their suppliers and who are investing in agroforestry and producer communities; and support companies using compostable packaging. Two brands I love are Alter Eco and GoodSAM.

Breakfast Cake with Preserve Swirl

Makes one 6-inch (15 cm) cake

We experimented for a long time as a team at west~bourne to craft a unique gluten-free muffin recipe. The original intention was to use cover crops that are essential to a regenerative farm ecosystem but are underutilized and often misunderstood, like buckwheat. With that North Star, we also challenged ourselves to create an all-natural gluten-free mix, free of chemicals and additive binders that often besiege these kinds of products. With that, it could be nourishing to the Earth and highly nutritious for us, even as a sweet finish to a meal.

Avocado oil, for greasing
1 cup (170 g) organic light
    coconut sugar
½ cup (80 g) gluten-free oat flour
½ cup (80 g) gluten-free light
    buckwheat flour
½ cup (80 g) gluten-free
    sorghum flour
2½ tablespoons psyllium husk powder
1½ tablespoons baking powder
1 teaspoon sea salt
2 teaspoons golden flax meal
2 teaspoons milled white
    chia seeds
1½ cups (355 ml) milk or unsweetened
    nondairy milk of choice
¼ cup fruit preserves of choice
Crème fraîche, whipped cream, or ice
    cream, for serving (optional)

Preheat the oven to 350°F (180°C). Line a 6-inch (15 cm) cake pan with parchment paper, lightly oil, and set aside.

Sift together the coconut sugar, oat flour, buckwheat flour, sorghum flour, psyllium husk powder, baking powder, salt, flax meal, and chia seeds. Gently whisk in the milk until combined and smooth, being careful not to overmix.

Pour the batter into the prepared pan and lightly tap the pan on the counter to settle the batter. Spoon the preserves onto the center of the batter and swirl it around using a fork for dispersion and design. Bake for 40 minutes, until a knife inserted in the center comes out clean. If it's not fully cooked, the cake will collapse out of the oven, so be sure not to take it out before it's ready. Remove from the oven and place on a wire rack to cool a bit. Serve warm with a dollop of crème fraîche, fresh whipped cream, or a scoop of ice cream. Or, just eat it straight out of the pan.

PRO TIPS: Transform this into a simple squash cake by replacing the preserves with ½ cup (127 g) pureed cooked squash or pumpkin, folded completely into the batter before cooking.

Seasonal Fruit Galette

Makes 6 servings

Pie is one of my all-time-favorite desserts. I remember when I was younger doing a road trip through the South with my family and tasting every single pie I could get my hands on. So, when we opened west~bourne, our restaurant in Soho, New York, I knew we had to have pie on the menu. In our teeny-tiny space on Sullivan Street, we didn't have the space or facilities to support any form of baking, so I went on a quest to find the best pie, which landed me squarely at Four and Twenty Blackbirds. It brought me right back to our bed and breakfast in Savannah, where I can still taste the peach pie I had for breakfast.

While pie requires quite a bit of skill to shape and trim, a galette is a free-form version where anything goes, so my hope is that this recipe will get you started on this baking journey. This galette will work with any fruit, though I like it with stone fruit, pears, apples, and berries the most. For the nut butter, I prefer pistachio or almond.

⅓ cup (40 g) whole-grain
  buckwheat flour

¼ cup (27 g) whole-grain oat flour

¼ cup (27 g) sorghum flour

1 teaspoon light coconut sugar,
  plus more for the crust

¼ teaspoon sea salt

⅔ cup (160 ml) avocado oil

1 cup (240 ml) water

1 large egg white, beaten

3 tablespoons smooth nut butter,
  at room temperature

4 or 5 ripe pluots

Ground nutmeg, cinnamon, or
  cardamom, for sprinkling (optional)

Combine the buckwheat flour, oat flour, sorghum flour, coconut sugar, salt, and oil in a large bowl and mix well. Add the water to the mixture, 1 tablespoon at a time, until the dough forms into a ball. One dough ball makes enough for two galettes; cut the ball in half and form each half into a disk. Wrap each disk in parchment paper and refrigerate for at least 1 hour and up to 2 days (you can also freeze one disk for up to 3 months, saving it for a future bake; thaw overnight in the refrigerator before using).

When ready to bake, preheat the oven to 350°F (180°C). Line a rimmed baking sheet with parchment paper.

Roll one dough disk between two sheets of parchment ⅛ to ¼ inch (3 to 6 mm) thick. Lay the free-form shape on the prepared baking sheet. Brush the egg white across the whole surface of the dough (this will seal it and prevent the bottom from getting soggy), saving a little for when you're ready to close up the galette. Let the egg white dry a little, about 2 minutes, then spread the nut butter across the dough, from the middle to the outside, leaving a 1-inch (2.5 cm) border untouched that will fold over the fruit and filling. This helps create a barrier between the fruit and the crust to prevent the moisture from the fruit from dissolving the crust, plus it adds a nice layer of fat, creaminess, and texture, and a slightly umami note.

Halve, pit, and thinly slice the pluots, keeping the slices attached at the side, then place the fruit over the nut butter skin-side-up, keeping a 1-inch (2.5 cm) border on all sides. Use your fingers to fold the edges of the crust over the pluots, leaving the center uncovered. Brush the crust with egg white again, and this is where you can sprinkle on any spice you might want to add.

Bake for 1 hour, until the crust starts to brown and crisp. Remove from the oven and cool on a wire rack before serving.

Fresh Fruit Bounty with Tahini Yogurt and Furikake Hazelnut Brittle

Makes 4 servings

The idea here is a Renaissance of fruit, not your average fruit plate. I encourage you to incorporate fruits you maybe have never worked with before, embracing the bounty our farmers work so hard to harvest. When we shot the photos for this book, I was inspired to try something new myself. I had kiwano melon for the first time in Paris when I dined at Table by Bruno Verjus. The first course is a thrilling exploration of fruit and vegetables locally and seasonally grown both raw and cooked and vibrant as an artist's palette. When you trim the spiky orange rind of the melon, the inner flesh and seeds are tart, bright, and subtly vegetal, adding an intriguing complexity. Sweet, savory, generous, and decadent, this dish is ideal at brunch or for a large-format table dessert.

The fruit for this recipe is flexible. Try it with passion fruit, grapes, stone fruit, apples, or pomegranate—anything fresh and in season is great. For the herbs, you can use fennel, basil, tarragon, or shiso.

1 cup (240 ml) plain, unsweetened yogurt
1 tablespoon tahini
2 teaspoons honey
Maldon salt
2 tablespoons finely chopped Furikake Hazelnut Brittle (page 225)
4 to 6 whole fresh fruit of choice (see headnote)
2 tablespoons extra-virgin avocado oil
2 tablespoons fresh herb leaves (see headnote)

In a small bowl, mix together the yogurt, tahini, honey, and a pinch of salt and set aside in the refrigerator. Put the brittle in a bowl in the freezer until ready to use. (If you don't have brittle, you can use any sort of seed-and-nut crunch you like, something that adds great texture.)

Prepare the fruit next, removing pits and cores as needed and slicing into large rustic pieces or even halves, depending on the size.

To plate, choose a large serving bowl or platter and spread the tahini yogurt over the bottom. Place the fruit on top of the yogurt, then garnish with the chopped brittle, a sprinkle of salt, a drizzle of avocado oil, and the herbs.

Granita Parfait

Makes 1 serving

I started thinking about this recipe because my kids are known for taking bites of fruit and then leaving behind a substantial amount of very good juicy bits. Of course those can be made into smoothies and ice pops or chopped up for snacks later, but I got to thinking about a low-sugar alternative to one of my favorite desserts: granita.

So, I started to play around with fresh-grating frozen fruit, which was a revelation. I now always keep some in the freezer along with our go-to ice cream so I can prepare what seems like a composed dessert for guests in relatively no time at all.

Sheep's milk yogurt, labneh, or ice cream
Frozen fruit (such as stone fruit, strawberries, or melon)
Extra-virgin avocado oil
Maldon salt

Place a scoop of yogurt in a champagne coupe, short glass, or small dessert bowl. With a Microplane, finely grate the frozen fruit directly over the yogurt, with enough to amply cover it in a mound, so the yogurt is a hidden surprise. Finish with a drizzle of avocado oil and pinch of salt.

Preserved Cherry Clafoutis

Makes 4 to 6 servings

Clafoutis is a lovely, curious cross between a custard and a cake, almost like a more sophisticated Dutch baby for the final bite of an evening. Though rustic, it's a showstopper, and you can't beat that smell when it's ready. It's meant to be eaten within an hour of coming out of the oven, so if you want a dessert that can be made in advance, head further into this section.

Though I love this clafoutis with preserved cherries, this is a great vessel for other preserved or fresh fruit you have on hand—ideally something that's a bit tart, such as berries, apricots, or grapes.

3 tablespoons avocado oil, plus more for greasing

½ cup (63 g) all-purpose flour

¼ cup + 2 tablespoons (61 g) light coconut sugar

1½ teaspoons kosher salt

3 large eggs

¾ cup (180 ml) whole milk

1 cup (230 g) pitted preserved cherries, drained

Unsweetened yogurt, labneh, or fresh whipped cream, for serving (optional)

Preheat the oven to 350°F (180°C). Grease a 9-inch (23 cm) cast-iron skillet or 1½-quart (1.4 L) round baking dish.

In a large bowl, whisk together the flour, coconut sugar, and salt. Whisk in the eggs and avocado oil. Add the milk and whisk until light and very smooth, about 3 minutes. Pour the batter into the prepared baking dish and scatter the fruit over the top. Bake for 45 to 55 minutes, until golden, puffed at the edges, and set all the way through. Let cool slightly, about 5 minutes, and serve alongside unsweetened yogurt, labneh, or fresh whipped cream, if desired.

PRO TIP: The preserved cherries can be homemade or store-bought, such as Luxardo or Fabbri.

Miso Chocolate Chip Cookies

Makes 20 cookies

Soybeans serve our soil as a notable cover crop. Miso is made by fermenting soybeans with salt and koji and is heralded as one of Japan's greatest superfoods. For this gluten-free riff on a classic cookie, the addition of miso brings a salty umami, almost like a caramel.

Chocolate chunks are preferred here, but chocolate chips or other small shapes will also work.

1½ cups (300 g) light coconut sugar
1 cup (240 ml) avocado oil
1 tablespoon white miso
2 tablespoons honey
2 teaspoons vanilla extract
2 large eggs
2¼ cups (210 g) oat flour
1 teaspoon baking soda
¼ teaspoon kosher salt
1½ cups (220 g) chopped
    dark chocolate
Flaky salt, for finishing

Cream the sugar, oil, and miso together in a stand mixer fitted with the paddle attachment on medium-high speed for about 2 minutes, until the mixture is the texture of wet sand. Scrape down the sides of the bowl, then add the honey and vanilla. Turn the speed to medium and mix for 1 minute, until combined. Beat in the eggs one at a time, until each one is incorporated. Scrape down the sides of the bowl again. Add the flour, baking soda, and salt and mix on low speed until starting to incorporate, then increase the speed to medium for about 1 minute, scraping down the sides of the bowl halfway through. Remove the bowl from the mixer and fold in the chocolate. Move the dough to the freezer (transferring to a smaller freezer-safe bowl as needed) for at least 2 hours, or ideally overnight, for the dough to rest and solidify. Cover if keeping overnight.

When you're ready to bake, preheat the oven to 350°F (180°C). Line two rimmed baking sheets with parchment paper.

When the oven comes to temperature, working quickly, scoop heaping tablespoons of dough, roll each ball between your hands, and place evenly on the prepared sheets, leaving 2 inches (5 cm) of space between each ball. Sprinkle the tops with flaky salt. If the dough is too firm to scoop, let it sit at room temperature for about 10 minutes; if the

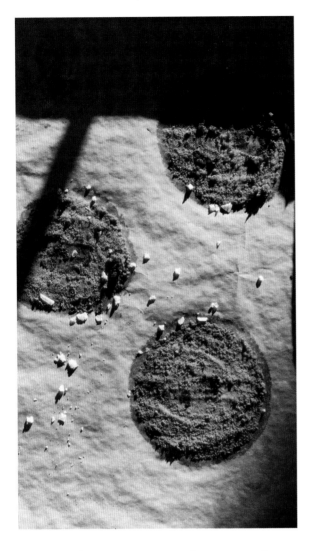

scooped balls are starting to melt, return to the freezer for 10 minutes before baking. Keep the remaining dough in the freezer.

Bake for 12 to 14 minutes, or until the edges of the cookies are golden brown. Let the cookies cool on the sheets for 10 minutes, then serve warm or move them to a rack to cool completely. While the cookies will still be soft when you remove them from the oven, they will harden to a perfect chewy consistency. Bake the remaining cookies in batches as desired. If you resist eating all the cookies immediately, you can store them in an airtight container at room temperature for up to 1 week.

PRO TIP: You can keep the dough in the freezer and bake individual cookies anytime you wish.

Brioche Doughnut Holes

Makes about 15

The legend of Stan's Donuts in Los Angeles will far outlive the restaurant's life, which sadly ended in recent years. Stan's was in the heart of Westwood and kept prices at 1950s levels despite massive growth in popularity. The air hung with warm, yeasty notes in the tiny shop, still a core memory I can conjure from regular family excursions.

A note on flour: There are so many sources for regenerative, climate-centered flours. From King Arthur to Central Milling to Local Millers, there are plenty of mindful options that I encourage you to try and hopefully integrate as your pantry staples.

Dough

8 cups (1.1 kg) all-purpose flour,
plus more for dusting

1 cup (160 g) light coconut sugar

2 tablespoons (30 g) kosher salt

1½ tablespoon (13 g) active dry yeast

1 cup (240 ml) milk

4 large eggs

½ cup (120 ml) avocado oil

Topping Mix

1 cup (130 g) light coconut sugar

2 teaspoons ground cardamom

1 teaspoon ground cinnamon

¼ teaspoon sea salt

For Frying

3 to 4 cups (710 to 945 ml) avocado oil

To Serve (optional)

Preserves, crème fraîche, and/or honey

To make the dough, combine the flour, sugar, kosher salt, and yeast in a stand mixer fitted with the dough hook. Add the milk and eggs and mix on medium speed for 10 to 12 minutes, scraping down the sides with a spatula as needed. Halfway through, add the avocado oil in a steady stream and continue mixing until the dough is well combined and starts to pull away from the sides of the bowl, about 5 minutes. Transfer the dough to an airtight container and let the dough ferment in the refrigerator for 24 hours.

The next day, transfer the dough to a clean work surface and cut the dough in half. Put one half back into the container and refrigerate to use at a later time (or freeze it for up to 1 month for future use; thaw on the counter overnight before using).

Roll the dough into a log 1½ to 2 inches (3 to 4 cm) thick, adding flour to the work surface as needed so the dough doesn't stick. Cut 1-inch (2.5 cm) portions using a small knife or bench scraper. The pieces should look like large gnocchi, plump and slightly squared with four corners. Transfer the dough pieces to a baking sheet until ready to fry.

To make the topping mix, in a large bowl, whisk together the sugar, cardamom, cinnamon, and sea salt.

When you're ready to fry, pour the oil into a large, wide pot and heat over medium-high heat until the oil reaches 350°F (180°C). Gently place up to 5 dough balls into the hot oil one at a time, give them a minute or so, then flip. When cooked and golden brown on all sides, use a slotted spoon to transfer them directly to the bowl of topping mix. Toss to coat, then serve hot or at room temperature as is or with the garnish of your choice.

Avocado Oil Cake

Makes one 9-by-5-inch (23 × 13 cm) loaf or 9-inch (23 cm) round

The best olive oil cake I've ever tried was made by Katherine Thompson, pastry chef and one of the owners of dell'anima, the first restaurant I worked at in New York City. Her cake brought me right back to my days of living in Rome. Only Katherine seemed to be able to craft that delicate balance of robust flavor, toeing the line between sweet and savory, with a crispy crust surrounding a pillowy, light-as-air sponge. When west~bourne started making avocado oil, I became obsessed with honoring one of my first chef mentors. Our version of this ancient Mediterranean cake, revered for its medicinal qualities, serves as a base recipe from which you can experiment once you master the finesse of baking it.

1 cup (240 ml) avocado oil, plus more
    for greasing
¼ cup (60 ml) milk or unsweetened
    nondairy milk of choice,
    at room temperature
½ teaspoon lemon juice or apple
    cider vinegar
1 cup (162 g) light coconut sugar
    or honey
3 large eggs
2 cups (250 g) all-purpose flour
1½ teaspoons baking powder
½ teaspoon kosher salt
¼ teaspoon baking soda
Crème fraîche (optional), for serving
½ cup Dukkah-Spiced Honey
    (recipe follows)

Preheat the oven to 350°F (180°C). Grease a 9-by-5-inch (23 × 13 cm) loaf pan or 9-inch (23 cm) round cake pan. For extra protection, you can line the bottom of the pan with parchment paper and grease the parchment. Chill the pan in the refrigerator while you make the batter (this will enhance the barrier and help your cake to not stick to the pan).

Combine the milk and lemon juice in a large bowl and let sit for 5 minutes to curdle. Whisk in the coconut sugar until dissolved, about 30 seconds, then whisk in the eggs one at a time. In a medium bowl, whisk together the flour, baking powder, salt, and baking soda, then gently whisk the dry ingredients into the wet ingredients. Using a spatula, fold in the avocado oil a little at a time (if you whisk in the oil, it might get too bitter), then transfer the batter to the chilled pan. Smooth the top of the batter with a spatula.

Bake for 45 to 55 minutes, or until a paring knife inserted in the center comes out clean. Let cool for 30 minutes before unmolding and slicing. Serve with a dollop of crème fraîche, if you like, and the dukkah-spiced honey.

Dukkah-Spiced Honey
Makes 1 cup (240 ml)

One more way to incorporate dukkah into your cooking is as a spiced honey. Simply combine ½ cup pistachio dukkah (page 48) and ½ cup honey, loosened slightly with a little hot water so it's pourable and can easily be mixed. The flavors get richer as they infuse together, so I like to jar this and have it at the ready. It's great for both savory and sweet dishes.

Carrot Cake with Passion Fruit and Chamomile Flowers

Makes one 9-inch (23 cm) cake

There is one thing that no one makes better than my mom: carrot cake. Since I can't match her on this one, I decided to marry her approach and secret ingredient (pineapple) with a cover-crop cake base so that it's lower in sugar, higher in fiber and vitamins, and naturally gluten free.

Carrot cake has always been nostalgic for me, and though it's old-school in so many ways, when I ask people what their favorite cake is, carrot cake comes up far more often than you'd expect. Too often recipes for it come out too dry, too dense, or with raisins—which is just a hard line for me. This one is fluffy and moist and gives that holiday warming with the addition of the spices. I love to decorate it with passion fruit and chamomile flowers (or even saved fennel fronds) for an added tart kick, slight crunch, and herbaceous bright finish, plus that way it looks like it was plucked right out of a garden.

This recipe is great for a proper tiered cake—like one you'd make for a wedding or birthday—but I've also made it into individual cakes for events and small dinner parties, which is a delightful surprise for guests when they receive their own personal cake. If you want to have a tiered cake (personally for this, I like three layers to really make it special), then you need to triple the recipe.

If you can't find light buckwheat flour, dark is OK as well, but it will affect the color of the cake. If you start with top-on carrots, you can save the leaves for vegetable stock. There no need to peel the carrots; just roughly grate them in a food processor.

1½ cups (360 ml) avocado oil, plus
   more for greasing
1 cup (170 g) organic light
   coconut sugar
1 cup (92 g) oat flour
1 cup (120 g) light buckwheat flour
1 cup (138 g) sorghum flour
1¼ tablespoons (22 g) psyllium husk
   powder
1¼ teaspoons (5 g) baking powder
1 teaspoon (4 g) sea salt
1½ teaspoons (4 g) golden flax meal
1½ teaspoons (4 g) white chia seeds,
   milled
2 teaspoons ground cinnamon
1 teaspoon ground nutmeg
1 teaspoon ground cardamom
1 cup (165 g) chopped pineapple
1½ cups (360 ml) milk or unsweetened
   nondairy milk of choice
2 cups (232 g) finely grated carrots
1½ cup (180 g) chopped walnuts
   (optional)
1 recipe Cashew Milk Frosting (recipe
   follows)
Passion fruit and chamomile flowers or
   fennel fronds, for garnish (optional)

Preheat the oven to 350°F (180°C). Line a 9-inch (23 cm) cake pan with parchment paper and lightly grease.

In a large bowl, sift together the coconut sugar, oat flour, buckwheat flour, sorghum flour, psyllium husk powder, baking powder, salt, flax meal, chia seeds, and spices. Blend the pineapple into a puree in a food processor. Gently whisk the milk, carrots, pineapple, oil, and walnuts (if desired) into the dry ingredients, in that order, until combined and smooth (be careful not to overmix).

Pour the batter into the prepared pan and lightly tap the pan on the counter to settle the batter. Bake for 40 minutes, or until a knife inserted in the center comes out clean. (The cake will collapse out of the oven if it's not fully cooked.) Remove from the oven and place on a wire rack until completely cool, at least 1 hour.

When the cake is completely cooled, frost and decorate as you wish.

Cashew Milk Frosting

Makes enough to frost a 9-inch (23 cm) cake

1½ cups (210 g) unsalted raw cashews
⅔ cup (160 ml) coconut yogurt
¼ cup (60 ml) maple syrup
¼ cup (60 ml) melted coconut oil
2½ teaspoons apple cider vinegar
1 teaspoon vanilla extract
¼ teaspoon grated lemon zest
¼ teaspoon sea salt

Soak the cashews in very hot water for 1 hour or in cool water for 6 to 8 hours, then drain. Put the cashews in a blender and add the yogurt, maple syrup, coconut oil, vinegar, vanilla, lemon zest, and salt. Blend, starting on low speed and then increasing to high, until smooth and creamy.

Transfer the frosting to a bowl, cover, and put in the freezer for 45 minutes. Chilling will help it to set and make it spreadable.

Remove the frosting from the freezer and gently whisk for 20 to 30 seconds. Cover the bowl and put it back into the freezer for another 45 minutes. Again, remove the frosting and whisk for 20 to 30 seconds, then return the bowl to the freezer and leave it there until you have a cheesecake-like consistency, about 3 hours.

Finally, remove the frosting from the freezer and blend until creamy and smooth again (if it's too thick to blend, let it thaw a little bit). The frosting keeps for up to 1 week in the refrigerator and about 1 month in the freezer.

Cream Cheese Frosting

Makes enough to frost a 9-inch (23 cm) cake

2 cups (240 g) powdered sugar
4 ounces (115 g) cream cheese,
    at room temperature
1 teaspoon fresh orange juice
½ teaspoon vanilla extract

Combine the powdered sugar, cream cheese, orange juice, and vanilla in a bowl and blend with a hand mixer (or use a food processor). Add drops of water, if needed, until it reaches your desired consistency.

Champagne Affogato

Makes 4 servings

Years ago, I was hired by a French luxury fragrance company to do a multicourse seated dinner for about thirty guests, where each course took them through tasting notes of the new fragrance line. The goal was to create an immersive experience around botanicals and scent. We presented this dessert as a fanciful Marie Antoinette–inspired play on a classic Italian affogato. Perhaps it has nothing to do with it, but I just love the interactivity of splashing something over ice cream and changing its shape, texture, and flavor. Like a lab experiment from high school but with high-end culinary ingredients. More than anything, it's just fun.

You can use any sparkling beverage for this, such as kombucha or sparkling zero-proof wine. And there's always vegan ice cream or sorbet if you prefer. The ground granola is there for a bite and crunch; you can use just about anything for this element, too—puffed rice, chia seeds, crumbled biscuit, or a crushed sweet snack mix.

½ cup (56 g) Sweet Toasty Seeded Granola (page 100)
Dried rose petals (optional)
Vanilla ice cream
Strawberries or cherries, stemmed or pitted, as needed (optional)
1 cup (240 ml) champagne

Grind the granola using a mortar and pestle, or pulse in a food processor until fine. You can add a few dried rose petals into the granola for a nice floral note, if you like.

Place a scoop of ice cream in a champagne coupe. Sprinkle a generous pinch (1 to 2 tablespoons) of the granola dust over the ice cream. If you want to add fruit, gently lay it around the ice cream at this stage. Repeat with the remaining ingredients.

Finally, tableside, pour a splash of champagne over each glass as your guests delight.

Yuzu (or Any Citrus) Posset

Makes 4 servings

You will wonder at the ease and science of this elegant dessert. Its origins are in old England, where a posset was a milk drink thickened with wine. It also reminds me of the Italian tradition of serving a fruit sorbet in its skin as a fresh finish to a meal. It's more special to keep the leaves and stem on the fruit, so it feels plucked right from the orchard onto your table.

The beauty of this dish is that you can prepare it in advance (the key to actually enjoyable hosting is knowing that some of the dishes and courses are already taken care of long before guests arrive). The key is to watch the milk fastidiously so it doesn't boil over.

For this recipe, you can mix and match any citrus, like orange or calamansi, but make sure if you are using anything other than yuzu or lemon that you combine it with some lemon or citric acid to increase the tartness and help the posset set.

2 to 3 yuzus or lemons
⅔ cup (165 ml) Cashew-Date Milk
    (page 53)
¼ cup (60 ml) honey
¼ cup (40 g) light coconut sugar
½ teaspoon kosher salt

To juice the fruit but retain the rinds, cut the citrus in half and scoop all the flesh out with a spoon. Squeeze the scooped-out fruit in a citrus press or with your hands into a strainer set over a small bowl. You will need 5 tablespoons (75 ml) juice. Set the rinds aside.

Combine the cashew-date milk, honey, and coconut sugar in a medium saucepan and bring to a gentle boil (be careful that it doesn't overboil) for 5 minutes. Remove from the heat, stir in the yuzu juice and salt, and let rest for 15 minutes.

Place the yuzu rinds on a plate and pour the custard evenly into each. (If you have more yuzu rinds than servings, use the rest for citrus dust [see page 47].) Put the possets in the refrigerator to set for at least 2 hours or up to 24 hours. This is best served cold.

Macerated Cherries, Fresh Cream, and Coconut Crumble

Makes 2 to 4 servings

Maceration is a simple technique, similar to marinating, of soaking fresh fruit in a liqueur, which imparts a new flavor and also softens and plumps the fruit. I like to pair somewhat alike but slightly different flavors with the fresh fruit and the liqueur. Specifically, I love doing this method with a blackcurrant liqueur called C. Cassis, which is made from lightly fermented New York blackcurrants with hints of cardamom, bay leaf, citrus rind, lemon verbena, and wild honey. This rare fruit was formerly banned to protect commercial pine logging but is now sustainably grown by small producers in the Hudson Valley.

There are over thirty varietals of cherries; any kind will do, but I prefer Rainier, Bing, or Tulare. I encourage you to try and test as many as you can.

The coconut crumble is something we keep on hand in an airtight jar in our pantry. It can be used like granola, but it's grain free.

1 cup (125 g) sweet cherries,
  pitted and halved
1 cup (240 ml) fruit liqueur

Coconut Crumble
1 cup (70 g) raw, unsweetened
  coconut chips
1 cup (70 g) shaved blanched hazelnuts
½ cup (120 ml) maple syrup
¼ cup (60 ml) unrefined coconut oil,
  melted and cooled
1 tablespoon kosher salt
1 teaspoon ground cinnamon

To Serve
1 cup (240 ml) cold heavy cream
Sea salt

Combine the cherries and liqueur in a medium bowl and let sit for at least 20 minutes.

Preheat the oven to 325°F (165°C). Line a rimmed baking sheet with parchment paper.

To make the crumble, combine the coconut chips, hazelnuts, maple syrup, coconut oil, kosher salt, and cinnamon in a medium bowl. Spread out the ingredients on the prepared baking sheet and bake for 10 to 15 minutes, until golden brown and toasty. Remove from the oven and let cool completely. Once cooled and hardened, break the pieces up into chunks (you can store extra crumble in an airtight container in a cool, dark place for up to 1 month).

Put the cream in a large metal bowl and whip using a whisk for about 10 minutes, until it forms soft peaks. Add a pinch of sea salt to finish.

Spoon the whipped cream into bowls. Drain the cherries from the liquid and divide them among the bowls, on top of the whipped cream, then sprinkle with the coconut crumble and serve.

PRO TIP: Strawberries, mulberries, or blackberries can be used in place of cherries.

Chocolate Cashew Pudding
à la Nammos

Makes 8 servings

Over twenty years ago, I dined at Nammos restaurant on the beach in Mykonos, Greece. Though much of my memory of that trip has faded, I can still taste the fluffy richness of the large-format chocolate mousse they served to complete the meal. Folded into a massive bowl with just a wooden cooking spoon, it was decadent yet casual, elegant while rustic. I hope you whip this up in your kitchen and invite a crew over to savor every spoonful, maybe even straight out of the bowl. You will need to start this dessert the day before you want to serve it.

6½ ounces (180 g) dark chocolate, chopped

6½ ounces (180 g) milk chocolate, chopped

4 cups (1 L) Cashew-Date Milk (page 53)

Set up a double boiler on your stovetop to melt the chocolate, or improvise one by setting a metal mixing bowl over a saucepan containing about 3 inches (7.5 cm) water, and bring to a boil. Put the dark and milk chocolate in the upper pan and swirl the chocolate with a silicone spatula just until fully melted.

In the meantime, heat the cashew-date milk in a medium saucepan to just below a simmer. When the chocolate has completely melted, pour the cashew-date milk into a blender and blend on low speed. As the milk is blending, slowly pour the melted chocolate into the vortex. Blend until completely incorporated, increasing the speed to medium once all the chocolate has been added. Pour the mixture into a large bowl, cover, and chill overnight.

To serve, transfer the mousse to a trifle bowl or large serving dish, or spoon into individual bowls.

Poached Quince, Hazelnuts, and Aged Goat Cheese

Makes 4 servings

I've been intrigued by quince since watching Rosie Perez answer the famed *Jeopardy!* question in the film *White Men Can't Jump*. At first glance, quince seems difficult to work with—just the challenge I love to rise to. Raw quince is tough, astringent, and inedible, yet it has an inherently alluring floral scent and a striking pink flesh. You know it's ripe when the skin is deep yellow without that slight green tinge; you can leave the peel on or remove it for this recipe, as you prefer. Prepared well, quince is a treat and surprise just patiently waiting to happen. Don't judge a book by its cover, as they say. I hope this will inspire you to get acquainted with what just might become your new favorite fruit.

That said, for those who want to stay in their lane, this same method can be used with pears, apples, plums, cherries, or peaches. Instead of the wine, you can use a sweet vinegar or fruit juice if you prefer.

As a version of a cheese-and-fruit platter, I like to plate this simply. For the goat cheese, choose something luxurious like Cypress Grove Humboldt Fog.

4 ripe quinces
Juice of 1 lemon
4 cups (945 ml) water
1½ cups (360 ml) red wine
½ cup (120 ml) honey
3 tablespoons orange blossom water
1 vanilla bean, split lengthwise
3 whole cloves
1 to 2 tablespoons avocado oil
4 ounces (115 g) aged goat cheese, sliced
¼ cup (30 g) hazelnuts, toasted, skins removed

If you wish, carefully peel the quinces from stem to root in fluid motions to keep the integrity of the fruit's shape. (The peels can be composted or dehydrated for cocktails or to add to your granola.) Place the fruit in a bowl, pour in enough cold water to submerge them, and add the lemon juice to prevent browning.

Combine the 4 cups (945 ml) water, wine, honey, and orange blossom water in a medium saucepan. Add the vanilla bean and cloves. Bring to a simmer, stirring until everything is dissolved. Remove the quinces from the soaking water and gently lower them into the liquid. Bring the poaching liquid to a boil, then reduce to a gentle simmer over low heat. Partially cover the saucepan so the liquid doesn't reduce too quickly before the quinces are cooked.

Simmer for 45 minutes, until the quinces are pink and tender (it depends on the fruit; sometimes it can take up to 2 hours to cook all the way through). (At this point, you can let the quinces cool in their poaching liquid, then refrigerate for up to 1 week; they will turn a deeper red color.)

If you like, use a slotted spoon to transfer the quinces to a bowl, then cook down the poaching liquid over medium heat until it's reduced to a loose syrup, about 10 minutes, and use it as a sauce for the dessert. You can also save the poaching liquid for just about anything (cocktails, yogurt, dressing, marinating).

Brush the quinces with the avocado oil. Arrange one quince per dessert plate with a slice of the cheese and a few toasted hazelnuts. Drizzle the poaching syrup over top, if using.

PRO TIPS: Optional additions to the poaching liquid include dried rose petals, rosemary sprigs, star anise, strips of orange peel, and/or cardamom pods.

For a variation, serve the quince with panna cotta or alongside sheep's milk or coconut yogurt with a little avocado oil and sea salt sprinkled on top. There's a beautiful dance between the cold and tang of the yogurt and the warming sweet and tart of the quince.

Charred Pears, Salt-Baked Beets, Yogurt, and Seeded Honey

Makes 2 servings

Bringing vegetables into desserts is a tactic I enjoy playing with, and this dessert evokes the woods for me. It's quite the surprise to guests to have that unexpected twist that is also nourishing for our health. The beets are intended to bring a soft earthiness to the grit, smoke, and sweetness of the charred pears. The tang of the yogurt balances out the spice and crunch of the honey. It's a symphony of opposites that become complements.

Anjou, Bosc, Concorde, and Seckel pears are best for grilling, as they keep their shape. For the yogurt, you can use coconut yogurt, sheep's milk yogurt, regular yogurt, or labneh.

3 small golden beets

3 to 4 cups (720 to 960 g) kosher salt

4 tablespoons (60 ml) avocado oil

2 pears

1 cup (240 ml) yogurt

2 tablespoons Dukkah-Spiced Honey (page 209)

Maldon salt

Preheat the oven to 425°F (220°C).

Scrub the beets and remove the greens (saving them for stock, salad, or pesto). Fill a roasting pan or cast-iron skillet with the kosher salt and bury the beets in it, covering them completely. Roast for 1 hour to 1 hour 15 minutes, until a small knife inserted into a beet meets no resistance and comes out clean.

Remove the pan from the oven and, using a spoon, carefully crack the baked kosher salt. Remove the beets and place them in a colander. Using gloves or a tea towel, carefully peel the beets while they are still hot (they will be easier to peel), gently rubbing the skin off and pulling it away from the flesh. (Alternatively, you can rinse the beets and let them cool to the touch, then peel them with a paring knife from the stem down.) Cut them in half lengthwise and toss in a medium bowl with 2 tablespoons of the avocado oil and a pinch of the baked kosher salt. Let rest for 5 to 10 minutes.

Heat a grill to high heat.

Halve the pears lengthwise, keeping the stem and core intact. With a melon baller or small spoon, gently scoop out the core just enough to remove the seeds. Brush the pears with 1 tablespoon avocado oil and season with 1 teaspoon baked kosher salt. Place the pears flesh-side-down on the grill and cook until charred, about 5 minutes, then flip and repeat on the skin side. The pears should be cooked through but not falling apart.

Divide the yogurt between two dishes, lay two halves of a pear on top of the yogurt, then place three halves of the beets around the pears. Drizzle the dukkah-spiced honey and remaining 1 tablespoon avocado oil over the top and finish with a pinch of Maldon salt.

PRO TIP: The beets can be roasted in advance and stored submerged in oil in an airtight container in the refrigerator for up to 1 week.

Furikake Hazelnut Brittle

Makes 3 cups (300 g)

When I was growing up, my parents' favorite festive treats were Dark Chocolate Bark (page 194) and brittle that they would store in the freezer to last long past the holiday season. Hazelnuts sing in any dessert, in my humble opinion, and are a more sustainable option than many other nuts. They are a low-maintenance perennial crop that store more carbon, use less water, and reduce soil erosion, plus they can grow on sloped land, making them quite resilient and a positive contributor to soil health. The combination of the sweet, woodsy hazelnuts and the umami crunchy furikake brings out the rich sweetness of the brittle.

1½ cups (300 g) light coconut sugar

⅓ cup (80 ml) honey

Kosher salt

3½ tablespoons water

¼ teaspoon baking soda

3½ tablespoons unsalted butter, at room temperature

¾ cup (105 g) hazelnuts, toasted and finely chopped

7 tablespoons (40 g) Furikake (page 52)

1 teaspoon Maldon salt

Line a rimmed baking sheet with parchment paper.

Combine the coconut sugar, honey, and a pinch of kosher salt in a medium saucepan. Add the water and mix until creamy. Bring up to a simmer over medium heat, stirring frequently with a heatproof spatula, until the sugar caramelizes to an amber brown and the temperature on a candy thermometer reads 300°F (150°C), 15 to 20 minutes. The caramel must reach 300°F (150°C) or the brittle will not be hard enough.

Working quickly, remove from the heat and sprinkle in the baking soda (the caramel will foam up slightly). Add the butter, continuing to stir. Once the caramel settles, stir in the hazelnuts and furikake.

Pour the brittle into the prepared pan. Using a silicone spatula, spread the caramel as thin (less than ⅛ inch / 3 mm thick) and as evenly as possible. Sprinkle with the Maldon salt.

Let cool for 15 minutes at room temperature. Transfer to the freezer for 1 hour, then break into small pieces using a mallet or heavy spoon. Store in an airtight container in the refrigerator for up to 2 weeks or in the freezer for up to 1 month.

The Butterfly Effect of Gathering

It takes a village to do anything. And it certainly takes a community to start and sustain a revolution. Gathering is a powerful act of sharing and caring for one another. Humans inherently exist in a collective, feeding off one another's energy and taking in the context that binds us. Coming together enlivens us, and every touch point in that experience is a layered dialogue between us and that flows through us.

Our world now is flooded with efficiency, automation, technology, and speed. We are too remote, and many of us are barely coping in the wake of Covid and the past years of intense isolation. Let's slow it down and mess it up. Sharing a meal is like playing a vinyl record with its quirk, texture, and soul, even more precious in the context of our modern world. I want to bring us back to the village, to remind us of the vibrancy and purpose that come from being together as well as the unfettered joy, unexpected success, and generous prosperity of convening and meeting new people. I want to be so progressive that we go back in time. Celebrate the analog, the simple, and the imperfect. Revel in the revolution of nature's time.

We did something quite differently in creating this book. I've long been involved in the school food movement, beginning a decade ago in New York, even before I had children of my own. My engagement in this cause was sparked by one staggering fact: The average US Department of Education budget per child is $1 per day for three meals, and in many cases these are the only meals those children will eat that day. What's more, there is a stranglehold of large corporate conglomerates on those school food contracts, and to no surprise many corporate suppliers are perpetuating agricultural practices that are instigating rather than assuaging the climate crisis. The justification is the very slim budget, but no testing or alternatives are being explored. It's just how it is, I've been told by many school administrators and friends alike.

It was in New York that I discovered the Edible Schoolyard Project, a beloved organization among chefs. Founded in 1995 by famed chef and activist Alice Waters, the Edible Schoolyard Project is a nonprofit organization dedicated to the transformation of public education by using organic school gardens, kitchens, and cafeterias to teach both academic subjects and the values of nourishment, stewardship, and community. At the time, I was working with Danny Meyer and Union Square Hospitality Group, and I started bringing our chefs to public schools in all five boroughs so they too could inspire change where it might matter the most: with our children. We did cooking demonstrations with schools and talked about the plants in each garden and how they contribute to an ecosystem and give healthy nutrients to our bodies. Regenerative agriculture is a movement that is addressing today and what will come. Sustainability starts at the source, and the ultimate source is the next generation during the formative years, where lifelong behaviors and preferences are originated. These are the seeds we must plant.

As Michael Pollan wrote, "Kids begin to learn about food in all its dimensions—as an edible medium of culture, science, ecology, and even social justice. The Edible Schoolyard is an eloquent and practical answer to some of the most pressing questions facing us as a society."

When I moved back to California, I reconnected with the Edible Schoolyard Project through its dynamic executive director, Ashley Rouse. We talked a lot about what's next for the organization and how to integrate regenerative education into schools through their gardens. We also agreed that we need to galvanize mothers to the movement, so that their children can be educated and inspired. Ashley asked if we could host a dinner together to convene a community of changemakers passionate about agriculture and its critical role in addressing the climate crisis and shaping the future for the next generation. We wanted to bring people together to dream audaciously, to share wild but deeply traditional ideas, and to challenge the way it's been done.

The timing was beautiful kismet, as I had been thinking about how we could capture the idea of bringing a village together mindfully and without waste. The images and spirit of that gathering are here on these pages. It was an immersive, intimate evening with growers, educators, activists, writers, and chefs—breaking bread, cooking over fire, and sharing the harvest at Zuma Canyon Orchids, a bold new regenerative farm just outside Los Angeles. We cooked from these recipes and used whatever produce was left over from the prior few days of the cookbook photo shoot. No stalk was left behind, all decor was foraged on site, tables were made from fallen trees on the property, and every component was integrative, served on the farm, under the stars.

This final moment of the book is an ode to Donna Hay. Knowing my deep passion for food and hosting, my mom gave me one of Hay's cookbooks when I first went off to college. I used it to host massive dinner parties throughout my time in college, building a community and a space to explore. In the book, Hay wove in curated moments and set menus throughout, a shortcut guide to the cookbook for those looking to get straight to the kernel of the matter. Her pointed practicality was both romantic and useful. This was reinforced by a dear friend who calls me every so often to ask how to put together what dishes with what course as she dives into a cookbook for something she's hosting. It dawned on me that while chefs might devour every chapter, every dish, every word of a cookbook and revel in the challenge of crafting a meal's progression, many find the sheer volume of recipes in a cookbook overwhelming to navigate and execute as a succinct menu for a gathering.

As a restaurant hospitalitarian at my core, please then, let me take care of you.

Impromptu dinner for two

Crispy Za'atar Socca with Shallot Labneh
(prepared as canapés, page 90)

Mushroom Larb Lettuce Cups (page 98)

Roasted Vegetable Tagine (page 170)

Macerated Cherries, Fresh Cream,
and Coconut Crumble (page 216)

Yuzu Vesper (page 67)

Weekend brunch for the crew

California Pozole Verde (page 174)

Corn Ribs with Queso Fresco
and Espelette Pepper (page 82)

Big Salad Energy
(page 135)

Chili Egg (page 76)

Breakfast Cake with
Preserve Swirl (page 196)

Fresh Fruit Bounty with Tahini Yogurt and
Furikake Hazelnut Brittle (page 200)

Matcha Pop (page 63)

Chamomile–Almond Milk Punch (page 68)

Breakfast family meal

Buckwheat Dutch Baby (page 104)

Tofu-Banana "Yogurt" with
Market Fruit and
Seed Cacao Crunch (page 106)

Sweet Toasty Seeded Granola (page 100)

Fresh Melon Ume Seltzer (page 61)

Gathering for supper

Big Salad Energy (page 135)

Maitake Mushroom Semolina Milanese
with Shallot-Herb Labneh (page 152)

Brothy Ditalini and Chickpeas (page 181)

Brioche Doughnut Holes (page 206)

Blood and Blackberry (page 58)

Late-night snack

Tartines (page 150)

Granita Parfait (page 202)

Village feast

Bouquet Lettuce Wraps (page 86)

Garden Focaccia (page 77)

Melon as Itself (page 118)

Eggplant Green Curry
with Cardamom Rice (page 189)

Charred Jimmy Nardello Peppers
with Crème Fraîche and
Herb Salad (page 92)

Wildfire Sweet Potatoes (page 158)

Dark Chocolate Bark
with Bee Pollen, Rose Petals,
and Pink Salt (page 194)

Carrot Cake
with Passion Fruit and
Chamomile Flowers (page 211)

Avocado Oil–Washed
Martini (page 64)

With Gratitude

Words to live by: "first optimism, secondly patience, third imagination, and fourthly, courage."

I have wanted to write a cookbook for over five years now. It's been a long, trying road, as it's an unbearably vulnerable place to pitch something you've created with nowhere to hide and open yourself up to judgment. As a chef and activist, I've dreamed of reaching into your homes and kitchens, yet it's been a daunting experience. I consider myself many things and have always had a generalist mindset, but I'm not sure I really think of myself as a writer. I knew I had to write this myself, start to finish, however hard or intimidating it would be—and it has been. This experience is like developing a new language; it's been humbling to be at the beginning of something knowing I don't know much. Giving myself a break as a novice, hopefully, I'll have earned space and consideration in your cooking and in your mindset toward food and how it's grown for our next generation. Thank you for letting me be here and for buying this book, supporting me and, more importantly, the essential regenerative revolution.

I've long felt this section of books should lead, not follow, as it takes a community of givers and creators who dedicate so much to bring a book to life. As with hospitality itself, it's a massive team effort. First, to my family, who gave me the core inspiration, open space, and thoughtful grace to throw myself into this project headfirst, knowing I had their love, understanding, and support every single step of the way. It's grueling and demanding to make a cookbook, let alone while trying to run a startup, be a partner, and nurture three toddlers. Being an early-stage entrepreneur and working mom is one of the most complicated roles I've navigated, and young children pull you in so many directions simultaneously. At times, this path can be filled with guilt, insecurity, and the utter messiness of never feeling like you do either role particularly well, and something always seems to give—or, let's be honest, completely fall apart. The timing of this book is quite raw and cosmic in and of itself, as writing it aligned with the year that we struggled intensely to get pregnant with our fourth child. It's as if the universe wanted this book that is dedicated to regenerating the Earth for the next generation to be born alongside the final piece of our family. In fact, we did the IVF implant that finally worked mere hours after I wrapped the intense multiday photo shoot for the book. I think I've never been so blissful and utterly exhausted in my life. I'd like to think the creativity sparks and intensity of purpose surrounded by this magnetic energy of beautiful humans opened a window for our baby to enter, and now the release and tour for the cookbook will be just weeks following our daughter's birth. Twin paths intertwined with every ounce of this kismet energy poured into these pages, both giving me gifts that I'm overwhelmed to receive.

To Ben Rosser, from the day we met over ten years ago and you said if I ever chose to do a cookbook—which at the time I laughed off as a seemingly near-to-impossible goal and long before I had given it serious consideration—you wanted to shoot it. The seed was planted by you from the outset. I could never have guessed when I cold DMed you on Instagram many moons ago that we would build such an enriching collaborative and creative bond. Shooting with you is a transportive experience, and I love seeing our world through your lens as only you can capture it. We share an intuitive dance, speaking without words and with innate trust—and I must give you credit for being willing to play jazz with me through our process, all without an assistant, tech, or formal stylist. You are one of the most talented Swiss Army knives I've ever met. We weren't sure a publisher would agree to a book shot exclusively on film, let alone from a first-time cookbook author and a photographer not primarily known for food. I'm proud that we paved a new visual road for ourselves, honoring the slow, analog, unexpected, and wild nature of the medium. We've always taken a chance on each other, willing to swim in new ponds we'd never explored before, and this right here, our first cookbook, is a testament to our being first-timers and bravely so.

There would not be a cookbook without the soulful April Valencia. April is one of those rare Renaissance humans who is as carefree and considered about every aspect of life—exceptional at everything she tries her hand at to where you would resent it if she were not such a generous soul. From recipe developing, to food and prop styling, to photography perspective, to emotional support every step of the way—you were one of the biggest guardian angels and anchors for me throughout. My gratitude knows no bounds.

A heartfelt thank-you to Joey Boujo, who was the conductor of everything, air traffic control to my unplanned, unstructured musings. I wish every author were able to benefit from your multidimensional talents as a focused chef, brilliant operational savant, culinary tool master, and captain that kept all the trains on track, on time, always with a bright vibe and a smile on your face.

The one and only Kristen Millar—thank you for bringing your creative genius and sharp perspective to the visual narrative of this entire project. As someone who has known me so well and for so long, I am grateful for your keen ability to translate and transform my most inner ideas and feelings into a form and story that breaks through. Your sharp eye and rigorous commitment to quality over everything added such distinct finesse to every page of this book.

A massive thank-you to Raquel Rodriguez, Chloe Walsh, and Emily Ferretti, who hustled hard to assist in absolutely every way, big and small, with full hearts. You picked up every ball juggled and jumped in to help with endless enthusiasm, care, and focus—no easy feat with such a packed shoot schedule.

To my dearest Clariss Saraf, who jumped into glam for every day of the shoot to make sure I felt my absolute best. Truth be told, I will never not be anxious about being in front of the camera, the very reason I always gravitated toward the back of house in restaurants and production. You buckled up for this chaotic ride, running around after me while I balanced so many roles at once across kitchens, farms, and fields. To Jessica Jones, who showed up with the most incredible makeup, braving the heat, dirt, and chaos of the farm with elegance and a crazy amount of laughter. I am also deeply grateful to Rosie Assoulin and Stella McCartney for graciously dressing me for the cookbook. I love working with female entrepreneurs and visionaries leading leaps in progress for their industries. You craft such vibrant clothing with such mindful craftsmanship and intention within every single detail. You both exemplify that the future is ours to shape and that creativity can thrive in concert with a climate-conscious mindset.

To my publisher, Chelsea Green, and in particular Rebecca Springer and Matthew Derr, who shared such wisdom and patience with me. Your commitment to a light carbon footprint in the printed word is inspiring, with every last detail considered.

To my entire family and village of friends, you push me to grow, dream big, and take chances.

Thanks to Sam Rogers, Kevin Augunas, Trevor Jahangard, and the entire Zuma Canyon Orchids team for welcoming our crew with such open arms into your farm to share your grounded practices. You are showing what is possible with regenerative farming and forging the future for us all. To the growers, the cultivators, the farmers who are brave radicals—you are the true heroes of this movement, day in and day out, honoring and refueling our soil, the foundation of our lives.

Index

and melon aguachile, 115, 120–21
pickled, 28
currants, 216
leeks with pistachio-sorrel gremolata and, 154–55
curry
eggplant, with cardamom rice,
189–90, 231
kabocha squash soup, 112–13
Cypress Grove Humboldt Fog, 220

D

dark chocolate bark with bee pollen, rose petals, and pink salt,
194–95, 231
dashi, vegan, 20, 37
in caramelized onions, 137
in mushroom oyakodon, 110
in mushrooms, tatsoi, and
rice, 172
dates
in almond milk, 71
in cashew milk, 53, 215, 218
in whole stalk or bulb salad, 131
dehydrated citrus peels, 47, 52
desserts, 193–225
avocado oil cake, 208–9
breakfast cake, 196, 230
brioche doughnut holes,
206–7, 231
carrot cake, 211–13, 231
champagne affogato, 214
cherries with cream and
coconut crumble, 216–17, 230
cherry clafoutis, 203
chocolate bark with
bee pollen, rose petals,
and salt, 194–95, 231
chocolate cashew pudding, 218
chocolate chip cookies, 204–5
fruit galette, 198–99
fruit with yogurt and hazelnut
furikake brittle, 200, 230
furikake hazelnut brittle, 200,
225, 230
granita parfait, 202, 231
pears with beets, yogurt, and
honey, 222–23
quince with hazelnuts and goat
cheese, 220–21

yuzu posset, 215
ditalini and chickpeas, 181, 231
Dinner at the Long Table, 229
Dirt Candy, 5
donabes, cooking with, 37, 94, 110
doughnut holes, brioche, 206–7, 231
drinks, 55–71
almond milk, 68, 71, 230
avocado margarita, 57
avocado oil–washed martini,
64, 231
Blood and Blackberry, 58, 231
lemonade spritz, 56
matcha pop, 63, 230
melon ume seltzer, 61, 230
yuzu Vesper, 67, 230
dukkah
honey spiced with, 208, 209, 223
pistachio, 48, 109, 145, 209
in savory galette, 96
dukkah-spiced honey, 209
avocado oil cake with, 208
pears, beets, and yogurt with, 223
Dutch baby, buckwheat, 104–5, 230
Dyer, Wayne, 22

E

Ecology Center (CA), 12, 13, 159
Edible Schoolyard Project, 228
Edison Grainery, 15
egg
chili, 76, 230
cured yolk in lentil risotto, 187
eggplant, 148
in curry with cardamom rice,
189–90, 231
as fries, 81
in katsu sandwiches, 148–49
eggshells and cartons in compost, 24
einkorn salad with squash, 119
erosion, 10, 14
escarole, 157
with pine nuts, Fiore Sardo and
herb oil, 157
Espelette peppers, 41, 47
corn with queso fresco and,
82–83, 230

F

farmers' markets, 7
fennel, 23, 130

in whole bulb salad, 130–31
fermented foods
garden focaccia, 77–79
garlic honey, 46
hot sauce, 38
lemon kosho, 34
preserved lemons, 30
and vinegar flavors, 34
fig leaves, in herb oil, 40
figs
in fruit butter, 32
stone fruit carpaccio with
mushrooms and, 123
Finley, Ron, 12, 13
Fiore Sardo cheese, escarole, pine
nuts, and herb oil with, 157
flaxseed, in pistachio dukkah, 48
flour, 206
buckwheat. *See* buckwheat flour
chickpea, 90
oat. *See* oat flour
sorghum. *See* sorghum flour
tempura, 75
Flour + Water, 15
flowers, edible
carrot cake with, 211–13, 231
chocolate bark, bee pollen, and
salt with, 194–95, 231
tomatoes and fruit salad with, 127
tomatoes in vinaigrette with, 138
Fly by Jing chili oil, 41, 76
focaccia, garden, 77–79, 231
food deserts, 12
food magazines, 5
food photography, 5–6, 228–29
food scraps. *See* leftovers and
food scraps
Four and Twenty Blackbirds, 198
French Culinary Institute, 4
Fresno peppers, 38
fried squash seeds, 159–61
fritters, vegetable scrap, 84–85
Frog Hollow Farmin (CA), 170
frosting
cashew milk, 213
cream cheese, 213
frozen foods, 20, 23, 35
avocado pits, 57
fruit
beets, yogurt, and honey with,
222–23

About the Author

Named one of *Fast Company*'s Most Creative People in Business, Camilla Marcus is a chef, entrepreneur, activist, and mother challenging conventional ideas about our food systems and environmental stewardship. Drawing from her California roots, she founded west~bourne in 2018 as New York City's first certified zero-waste restaurant—keeping 91% of its waste out of landfills or incinerators and earning it gold-level certification by TRUE. After closing the restaurant's doors as a result of the pandemic, Camilla has expanded west~bourne's mission of eating well and doing better into a collection of regenerative, carbon-neutral provisions for the modern home.

Committed to cross-industry innovation, Camilla ranges in her endeavors from championing regenerative farming to reshaping the hospitality landscape. She co-founded ROAR (Restaurants Organizing Advocating Rebuilding) and the IRC (Independent Restaurant Coalition), which helped to pass legislation for key federal aid and raised more than $3 million for cash assistance to restaurant workers. Her passion extends to crafting a more equitable childcare system for working mothers. Camilla is also a member of the Fast Company Impact Council and part of the invitation-only Google Food Lab think tank.

Her work, alongside her marvelous "accidentally vegetarian" recipes and her forward-thinking philosophies, has been featured in a wide range of media, including the *New York Times*, *Food & Wine*, *Forbes*, *Vogue*, CNN, *Bon Appétit*, *Women's Wear Daily*, *Marie Claire*, *InStyle*, *New York* magazine, *Dwell*, *Nylon*, *Thrillist*, *Eater*, and many more. Through all her efforts, Camilla remains dedicated to fostering a community-minded, sustainable approach to eating, gathering, and planetary care.